TOXIC AIRS

TOXIC AIRS

BODY, PLACE, PLANET IN HISTORICAL PERSPECTIVE

EDITED BY
JAMES RODGER FLEMING
AND
ANN JOHNSON

UNIVERSITY OF PITTSBURGH PRESS

Published by the University of Pittsburgh Press, Pittsburgh, Pa., 15260
Copyright © 2014, University of Pittsburgh Press
All rights reserved
Manufactured in the United States of America
Printed on acid-free paper
10 9 8 7 6 5 4 3 2 1

Library of Congress Cataloging-in-Publication Data

Toxic Airs: Body, Place, Planet in Historical Perspective /
edited by James Rodger Fleming and Ann Johnson.
pages cm
Includes bibliographical references and index.
ISBN 978-0-8229-6290-8 (paperback : acid-free paper)
1. Air—Pollution—History. 2. Environmental sciences—History. 3. Science—History.
I. Fleming, James Rodger. II. Johnson, Ann, 1965–
TD883.T66 2014
363.739'2—dc23   2014001171

CONTENTS

ACKNOWLEDGMENTS

We thank the Chemical Heritage Foundation, especially Ronald Brashear, Jody Roberts, and Carin Berkowitz for their support and gracious hosting of the thirteenth annual Gordon Cain Conference, "Chemical Weather and Chemical Climate: Body, Place, Planet in Historical Perspective," March 31–April 2, 2011, where these chapters were first presented. Paper writers and discussants included Brenda Gardenour Walter, Marilyn Gaull, Christopher Hamlin, Ana Carden-Coyne, Susie Kilshaw, Roger Eardley-Pryor, Melanie Kiechle, Vladimir Jankovic, Peter Brimblecombe, Ann Johnson, Richard Chase Dunn, Jongmin Lee, Rachel Rothschild, David DeVorkin, Dania Achermann, Alison Kenner, Andrea Polli, Linda Richards, Matthias Dörries, Noah Bonnheim, and Jim Fleming. Rapporteurs Licia Peck, Nicole Sintetos, Victoria Feng, Roger Turner, Noah Bonnheim, Erin Love, James Bergman, Richard Chase Dunn, and Amy Fisher kept track of the discussions and provided helpful notes. Gregory Good, John Cloud, and Roy Goodman also provided valuable insights. A special note of thanks goes to conference discussants David Cantor, Roger Launius, and Marc Rothenberg, who provided commentary and led crosscutting sessions at the end of each day. All contributed to robust discussions and interactions at the meeting.

INTRODUCTION

THIS EDITED VOLUME examines toxic airs from the Middle Ages to the recent past on a variety of scales, from lungs to locales and from places to planetary processes. Chapters shed new light on the myriad ways that humans have feared and then made sense of the air they breathe and the climates in which they live. Such understandings have often opened avenues for intervention in the air, improving it for ourselves and sometimes making it more frightening for perceived enemies. The authors examine how cultural assumptions, technologies, and policies were formulated, often in an atmosphere of crisis, and typically aimed at mitigating fears about health, air quality, social tensions, and the global atmosphere.

The history of the air, and its woes, is a topic that falls among many different disciplines. While the composition of the air is ostensibly a problem of chemistry, chemistry itself has changed significantly since the Middle Ages. Even the notion of atmospheric chemistry is relatively recent and it raises the specter of presentism to see any kind of modern chemistry in medieval concerns about breath or in nineteenth-century concerns about climate and disease transmission. On the other hand, environmental history offers some analytical frameworks to analyze pollution, but typically focuses on terrestrial rather than atmospheric concerns. In addition the strong declensionist narrative of environmental historians is too simplistic for the complexities of toxic air, which is often surprisingly remediable. Medical histories of the role of air in human health address some of the concerns here but our scale transcends the body, to involve ecosystems, upper layers of the atmosphere that humans do not inhabit, and ultimately the health of the planet. Lastly, this volume points out that policy matters, and the history of air quality policies has to be seen in a context of available science and technologies, political possibilities, and emotional responses to the air. Humans are, not surprisingly, threatened by compromises to their air, and they have reacted by wielding their full arsenal of understandings on toxic airs.

This volume is at once evocative and visceral. It introduces a new, interdisciplinary set of voices from different specialties. The volume presents a broad series of case studies that, taken together, open up new conversations on a variety of scales of analysis (body, place, planet), while shifting perspectives away from discovery, progress, simple solutions, or gloom and doom. The delicate balance is to introduce the apprehensions, problems, and looming disasters involving the atmosphere while at the same time contextualizing the issues and connecting science, engineering, and medicine with public understandings, policies, and practices. The authors use multiple historical accounts of toxic air as lenses to view the connections between different ways of knowing and acting, focusing on historical actors' efforts to do something about what they perceive to be dangerous in the air they breathe or that threatens them from above.

The authors' intent is to understand many different perceptions of both air and the possibility of action regarding the air. To do so they employ a large array of disciplinary approaches at the intersections of history, philosophy, anthropology, literature, art, science, engineering, and policy. Coverage ranges from an analysis of toxic *pneuma* in the Middle Ages to current, cutting-edge policy concerns involving transboundary and global air pollution, from medical to bureaucratic and technoscientific concerns. In all cases, the chapters inform and are informed by one another as they speak to both specific and universal issues. The essays are empowering, providing new analytic tools and clear visions of the problems on the one hand and providing critiques of efforts to manage them on the other.

The authors intentionally vex technological fixes in hope that higher levels of solutions might emerge. The volume is designed to act as a catalyst increasing the speed of insights by bringing together new materials that previously have been separated by discipline or time period. The authors present a rich and diverse set of cases and introduce new categories that challenge perspectives. Taken together these essays facilitate new ways of imagining and talking about questions of the atmosphere and toxic air. They produce new insights precisely because of their breadth, range, and temporal and topical diversity.

Overarching themes addressed in more than one chapter include "toxic *pneuma*"; private and public guilt; the "pathology of deadly air"; the natural and the artificial; the visible and the invisible; the social organization of science, medicine, and technology; transmission of technical information to the public; technological "fixes"; the many challenges inherent in regulating the atmosphere; anxiety about personal and environmental health; and scale interactions along the axis of body, place, planet.

The volume begins on a personal scale, examining the ways that airs act as conduits between human bodies and contaminants. Chapters here focus on the intimate level of breathing, choking, disease, invisible wounds, and scars. They introduce the themes of hidden aerial threats, bodily permeability, fear, and vulnerability. Brenda Gardenour Walter takes as her focus the toxic breath of witches as exemplary of the centrality of air in the reciprocal relationship between body and environment, microcosm and macrocosm, in medical theory and theology as received from ancient traditions and modified by scholars in the thirteenth and fourteenth centuries. She examines the ways these seemingly dichotomous traditions inform one another and inform our basic assumptions moving forward. This chapter sets the stage for the rest of the volume, specifically in its claim that the body is not bounded mechanically, chemically, or pneumatically, but indeed is open to fluid and foul atmospheric influences. Her work stands as a corrective to the popular but misguided notion of the deep past as static and unchanging in contrast to the notion that our own times are the only ones changing dynamically at an increasing rate. Further, her work speaks to issues raised by other authors in this volume, including the permeability of the body and its vulnerability to invisible toxic airs released in warfare (Kilshaw) or in nature (Fleming).

Christopher Hamlin examines cholera research in nineteenth-century colonial India by regimental surgeon Reginald Orton and like-minded successors as pioneering but failed attempts to make a complete geographical, atmospheric, and etiological study of the disease. This study illuminates the social organization of medicine and empire and examines epistemological tensions between metropolitan and colonial accounts of disease and between empirical and rational research traditions. It also illuminates the modern quest to establish a synthesis of infectious and environmental factors involving disease. In this way Hamlin's paper informs the work by Kilshaw on Gulf War syndrome and the work of Lee on the Environmental Protection Agency's (EPA) Community Health Environmental Surveillance System (CHESS).

Is it better to cry than die? Roger Eardley-Pryor provides multiple perspectives on the use of tear gas as a technological fix in the Vietnam era, both in warfare and in crowd control. The paradox of nonlethal chemical agents lies in the fact that their use in combat was prohibited internationally, but their domestic use was widespread and considered humane. Policy changes were not driven purely by rational choices, however, but by the swirl of events and perceptions. Eardley-Pryor's focus on the elusiveness of a technological fix echoes through other chapters—for example, those by Brimblecombe and by Dunn and Johnson on urban air pollution and automotive emissions.

By design, tear gas intervenes between body and place, whereas the collateral damage to the atmosphere of Kuwait triggered the phenomenon known among veterans as Gulf War syndrome. Susie Kilshaw, an anthropologist, explores the fear of exposure to unseen agents and the uncertainty that followed the veterans home. In the aftermath of the war, and in the absence of a confirmed medical diagnosis, theories of causation began to revolve more closely around veterans' corporeal experiences and contested memories. Her work raises the themes of invisibility and subjectivity that resonate throughout the volume.

Air pollution, which is neither invisible nor subjective, also intervenes between body and place. Los Angeles air plays a central role in atmospheric chemist Peter Brimblecombe's account of the development of the photochemistry of smog. He helps us understand how chemists deciphered the complex relationships between emitted pollutants and the atmospheric processes that transform them into LA's infamous smog. He also argues that without an understanding of atmospheric chemistry it is difficult to develop effective policy.

The related chapter by Richard Chase Dunn and Ann Johnson explores the sequence of technological fixes developed prior to 1980 to curtail automotive emissions. This work, conducted in automotive laboratories and in policy arenas, is framed by studies that model and measure air pollution. Engineers incorporated complex chemical understandings of pollution and reframed that knowledge into design parameters, transforming air pollution into a manageable emissions problem. Such work also shaped and was shaped by new regulations governing the automobile.

For many activists, politicians, and intellectuals the establishment of the EPA was a major accomplishment in itself. However, scientists, engineers, and administrators faced several challenges in defining an agenda and delineating the boundaries of research and regulation. The chapter by Jongmin Lee presents the case of CHESS—a comprehensive epidemiological study of the effects of air pollutants in eight communities on the health of children, asthmatics, and the elderly—to show how EPA scientists and engineers dealt with new geographical, bureaucratic, and functional challenges. Lee's study employs the notion of coproduction to explain what the EPA expected from CHESS and why this flagship program fell short of those expectations.

While CHESS spanned the nation, E. Jerry Jessee's essay reaches into the stratosphere to unpack the controversy over atmospheric nuclear weapons testing and fallout. Was the fallout ultimately dissipated in the environment? Research on the movement of radiation through the atmosphere and biosphere indicated that it was not, nor was it benign. Such work played a critical role in formulating estimates of fallout risks by pointing to the complex and dynamic ways that the environment both modulated and magnified human exposure to fallout.

After the Limited Nuclear Test Ban Treaty of 1963, concerns about transboundary air pollution took center stage. Rachel Rothschild's chapter examines the Organisation for Economic Cooperation and Development (OECD) project to measure sulfur dioxide transfer between European nations during the 1970s; in doing so she explores the interaction of scientific research, international law, and national policies. The OECD project confirmed a significant transfer of acid rain precursors from northwest Europe to Scandinavia, and nearly two decades of ecological studies provided ample evidence of the environmental harms resulting from this pollution.

Also during the early 1970s, perceptions of ozone underwent a profound transformation. Previously regarded primarily as an annoying component of urban pollution, it began to appear as a positive element protecting the earth from harmful radiation, exemplified in new terms like "ozone shield," which often replaced the more neutral "ozone layer." Matthias Dörries details how new awareness of the role of the atmosphere in supporting life on earth brought the atmospheric sciences to the forefront of political and public debates. Some scientists were particularly concerned about the effect of a nuclear war on the ozone layer; others focused on the pollution load from a proposed fleet of supersonic transports. Dörries's chapter traces the evolution of the debate about ozone and its threats.

The current crisis in climate change and the realization that humans are the primary cause of this change has raised questions about ownership and responsibility. Who "owns" the crisis and who is responsible for mitigating and reversing it, if possible? The overwhelming response has been to propose a market solution. Andrea Polli, a conceptual artist, argues that the arts have a role to play. Her essay brings contemporary artists firmly into the conversation as it explores the idea of air for sale, discusses several contemporary art projects, and highlights the evolving role of artists in affecting public participation on climate change.

The personal and the planetary meet in James Rodger Fleming's chapter on carbon dioxide as toxic *pneuma*. He recounts the history of understanding the substance that eventually came to be known as CO_2, takes us on a Cook's tour of caves and valleys filled with the asphyxiating gas, and brings the story of carbon dioxide as an active agent in climate change up to date.

The authors in this volume have provided their best insights and inspirations on toxic airs over a vast range of spatial and temporal scales. They have also inspired one another. These essays provide both insights and challenges about the role of air and the atmosphere in the history of science and environmental history. We trust this volume will introduce new themes, provide new perspectives, and address new problems (perhaps frontiers?) in the medical, chemical, and environmental history of the atmosphere. We offer *Toxic Airs* as a model of interdisciplinary historical work in a number of genres and on a wide range of scales.

TOXIC AIRS

1

CORRUPT AIR, POISONOUS PLACES, AND THE TOXIC BREATH OF WITCHES IN LATE MEDIEVAL MEDICINE AND THEOLOGY

Brenda Gardenour Walter

"FAIR IS FOUL, and foul is fair, / Hover through the fog and filthy air."[1] So chant the Weird Sisters in the first act of Shakespeare's *Macbeth* as they stare into the vaporous gloom, gleaning premonitions of horrific events yet to unfold. Although appearing only sporadically, the witches drive the play's narrative, much as the winds they command drive sailors and their ships at sea to be "tempest-tost."[2] The parallel between the haggard and foreboding witch and the violent and pestilent wind is a commonplace found in many of Shakespeare's sources. Perhaps the most famous is Holinshed's *Chronicles of England, Scotland, and Ireland* (1577), in which the Weird Sisters are described as "three women in strange and wild apparell, resembling creatures of elder world," and "endued with knowledge of prophecy by their necromantical science."[3] Holinshed attributes the witchy woman's command over the natural world to her mastery of arcane wisdom; other sources contemporary with Shakespeare, however, paint a far darker portrait of the witch, a wicked and poisonous hag whose maleficent powers stem from her diabolical pact with Satan himself. In his *Daemonologie* (1597), James I of Scotland describes the witch as an evil creature who can suddenly and violently "rayse stormes and tempests in the aire, either upon Sea or land," an act made "verie possible" because of her master Satan's "affinitie with the aire as being a spirite, and having such power of the forming and mooving thereof."[4] Martin Castañega's *Treatise on Witchcraft* (1529) and Reginald Scot's *Discoverie of Witchcraft* (1584), both of which offer a far more skeptical treatment on the topic, nevertheless served to reinforce deeply embedded and longstanding cultural associations between the witch and the "foul and filthy air"—whether she be riding upon it, directing it, or poisoning it with her own corrupt breath.

The strange union of witch and pestilent air found in the witchcraft treatises and concomitant witchcraft culture that informed Shakespeare's Weird Sisters was not a creation of the early modern period. Instead, the foundations for the sixteenth- and seventeenth-century Satanic witch, with her corrupt and melancholic body, poisonous breath, and command of the air, can be located in the academic milieu of the thirteenth and fourteenth centuries, where theologians and physicians shared a language of discourse and a set of epistemologies rooted in the logic and natural philosophy of Aristotle (ca. 350 BCE).[5] Aristotle's *Logic* as well as his *Prior Analytics* and *Posterior Analytics* provided scholars with rhetorical tools for theoretical investigation and argumentative proof of both supernatural and natural phenomena, while his *Libri Naturales*, including the *Physics, On the Heavens, Meteorology*, and *On the Soul* provided them with a framework of natural law by which the behavior of creation was bound.[6] For medieval natural philosophers, physicians, and theologians, the air in all of its expressions was a topic of critical investigation because of its unique properties. One of the four theoretical and therefore imperceptible elements that formed all matter, air might also physically manifest in tangible, if still mostly invisible, forms such as mild and tempestuous winds as well as fair and foul breath. Air could ascend to the ethereal realms to mingle with the divine, then descend to earth to permeate and change the very nature of the unbounded medieval body. Because of its fluidity, its invisibility, and its association with the supernatural, the air—be it the atmosphere, wind, or breath—was ascribed great power.

The amorphous and transformative power of air was the subject of discourse for Aristotle, Hippocrates (ca. 400 BCE), and Galen (ca. 150 CE), the latter of whose authoritative works served as the foundation for learned medicine in the medieval university. Galen's theory of *pneuma*, in which air taken in through the lungs becomes imbued with vital and psychic spirits, both provided air with an increased role in human physiology and the etiology of disease and linked internal bodily air with the spiritual realm. Galen's *pneuma*, when conflated with the soul, provided scholastic theologians versed in medical theory with one more discursive tool in the theological construction of the human and superhuman body. Under the quills of clerics in the thirteenth and fourteenth centuries, Aristotelian natural philosophy and learned medicine became handmaidens to patristic sources (the authoritative works of the church fathers), hagiography, and folk traditions in the creation of two imaginary bodies: that of the holy virgin, whose *pneuma* was pure and health-giving, and that of her categorical inversion, the evil woman, whose *pneuma* was corrupt and pestilent, foul and filthy. The construction of these two categories would shape later medieval witchcraft treatises such as the *Formicarius* (1435–37) of Johannes Nider and the *Malleus Maleficarum* (1486) of Henry Kramer and Jacob Sprenger, both of which perpetuated the connection, promulgated through centuries of scholastic argumentation, between foul air, noxious humors, and the pestilent body of the corrupt and wicked woman. In turn, these treatises and the culture of persecution that supported them inspired the witch hunts of the early modern period, fueled James I's accusations that North Berwick witches were behind the demon-laden winds threatening to sink his ships, and made plausible Shakespeare's Weird Sisters and their command over the pestilent air.

ARISTOTLE AND THE MEDIEVAL SCHOLASTICS

Medieval medical and theological conceptions of the amorphous air and its reciprocal relationship with the local environment, the body, and the spiritual realm are rooted in the academic study of natural philosophy, at the heart of which lay the ancient cosmos, the closed system within which the laws of physics and metaphysics might function. It was this cosmological structure that provided both physicians and theologians with a theoretical framework upon which to construct their understanding of the natural and supernatural worlds. The cosmos itself was composed of several concentric crystalline spheres, nesting like matryoshka dolls. Beyond the outermost sidereal sphere lay the realm of Aristotle's prime mover,[7] understood by medieval natural philosophers as the God of Christianity, who created the cosmos and set it into motion out of love, while below the sidereal sphere lay those of the planets, ever-spinning in perfect harmony through the fifth element, aether. In this realm above the moon, all movement was circular and eternal, and change—in any form—could not exist. Below the moon, however, the cosmos became a very different place, a world of chaotic and often violent motion. Following Empedocles and Plato, Aristotle holds that the world below the moon is composed of four theoretical elements: fire, air, water, and earth, each of which has its own set of qualities and natural place of rest.[8] This hierarchy describes each element's place of rest according to its assigned qualities; while the elements theoretically desire to be in their natural places of rest, in actuality they rarely achieve them because of the continual motion and change that predominate in the chaotic sublunary realm. At the center of the cosmos lies cold and dry earth, the most corrupt and unstable and therefore the heaviest of all elements, upon which runs elemental water, cold and wet. Above earth and water rise the lighter elements: air, which is hot and moist and extends into the sky far beyond human vision, and fire, which is hot and dry and burns in a fiery ring at the outermost perimeter of the earth's atmosphere.[9] It is here that the lightest element on earth—fire—comes into contact with the ethereal realm of the divine. Aristotelian aether (unlike the mundane aethers of Anaximenes, Empedocles, or Plato) was perfect and eternal in nature. It could not actively interact with the earthly elements, but it did come into physical contact with the ring of fire that brushed along its underbelly.[10] Here, at the boundary, the atmosphere of the divine celestial realm and that of the earthly realm made a connection that facilitated the conceptualization of an elemental gradient that rose from heavy, dross earth to lightest air and fire and ultimately to divine aether. For later medieval theologians, this atmospheric matrix of aether, fire, and air would become the conduit through which moved celestial beings, such as angels, divine spirits, and even demonic entities.

Medieval natural philosophy was concerned not only with abstract and elemental air, which might be found in different proportions in all earthly matter, but also with atmospheric air made manifest in meteorological phenomena, including rain, clouds, and wind.[11] Aristotle holds that atmospheric air "surrounding the earth is moist and warm, because it contains both vapor and a dry exhalation from the earth"; as the air condenses it produces clouds and as it cools, it condenses once "again and turns from air into water. And after the water has formed it falls down again to the earth" in the form of rain.[12] According to *Meteorology*, air circulates not only vertically because of continual material transformation and the laws of sublunary physics but also horizontally because of the spherical rotation of the prime mover, for "it is the revolution of the heaven which carries the air with it and causes its circular motion."[13] Once set into motion, however, atmospheric air does not maintain its circular path, but instead moves erratically across the face of the earth, a premise proven by the actions of the wind. Aristotle argues that terrestrial winds are not merely atmospheric air in motion, but "earthy exhalations" that rise from the four cardinal directions, with the north wind, known as *Septentrio*, being cold and dry; followed by the south wind, or *Meridies*, which is hot and wet; the east wind, *Oriens*, which is cold and wet; and *Occidens*, the west wind, which is hot and dry.[14] Just as popular were theories culled from medieval handbooks that drew on sources such as Pliny the Elder and Isidore of Seville, which argue that the winds might be stored in caves, rise from the sea, or be stirred by the effects of the sun, moon, and planets.[15] Between each of the cardinal winds blow intermediate winds, for a total of twelve. The qualities of these winds, however, are less than clear. It would seem that two distinct models were at work in determining the qualities of the cardinal winds. In *Airs, Waters, and Places*, Hippocrates asserts that the north wind is cold and dry, while the south wind is cold and moist, which contradicts the *mundus-annus-homo* (world-year-human) model depicted in the *De Rerum Natura* of Isidore of Seville and embraced by medieval academic medicine.[16]

For medieval theologians, the long-distant heirs of Aristotle, the winds had a multifactorial genesis and were as much supernatural as natural—laden with unseen forces. The cardinal winds, for example, were named after the Greek *Anemoi*, or wind gods, who directed the gales according to their temperament. In medieval manuscripts, *Septentrio*, so named after the seven-star constellation associated with the north sky, is also known as *Boreas*, Greek god of the north wind whose name translates as "the devouring one," a reference to his biting breath as well as his violent and unpredictable temper.[17] Medieval wind diagrams reveal another, less obvious, link to the supernatural world of antiquity—the belief that the winds carry demons. In the ancient world, the wind was believed to be teeming with spirits that might be good, bad, or neutral. Medieval Christian theology, bound by an Aristotelian contrariety that did not allow for neutral spirits, placed angelic beings in the divine realm beyond the moon, and condemned evil demons to the melancholic winds of the earth. In wind diagrams, the demons of the air were often drawn as foul, dark, winged creatures that were both the source and substance of the tempestuous winds that flew from their mouths.[18]

At the center of the cosmos, open to the swirling air and irrational winds, lay the human body, a microcosm that contained "something of fire, of air, of water, and of earth."[19] According to medieval medical theory, the body was composed not only of the four elements but also of the four humors—blood, yellow bile, black bile, and phlegm—each of which was assigned a set of Aristotelian qualities that corresponded with a season of the year and an element. Blood was hot and moist, allied with springtime and elemental air, while yellow bile was hot and dry, and linked with summer and fire. Black

FIG. 1.1. Author's line drawing and translation of the "Mundus, Annus, Homo" from Isidore of Seville, *De Rerum Natura*.

bile was cold and dry, the humor of autumn and of elemental earth, while phlegm was cold and moist, and linked with both winter and water. The bodily humors were likewise allied with specific organs, phases of the moon, zodiacal signs, regions, and cardinal directions.[20] The seamless integration of the human body with the macrocosmic system that lay at the foundation of humoral theory was represented visually in *mundus-annus-homo* roundels (see fig. 1.1) such as that contained in Isidore of Seville's *De Natura Rerum*[21] (ca. 600 CE) and elaborated upon in medieval Latin medical texts such as the *Viaticum, Pantegni, Articella, Almansoris, Qanun*, and *Lilium Medicinae*.[22] As part of a larger gradient, the medieval body was contiguous with sublunary and superlunary forces that could disrupt a healthy humoral balance, a state unique to each individual and almost impossible to maintain, thereby causing illness. For example, too much red wine, which was hot and wet, consumed in the hot and wet springtime would engender the production of an abundance of hot, wet blood, causing fever, excessive sweating, sleeplessness, and an increase in libido. The therapy prescribed for such a condition included venesection in order to purge the body of excess blood, a restriction on hot and wet foods such as red meat and wine, and the prescription of a qualitatively cold and dry diet in addition to cool baths and calm music.

HIPPOCRATES, GALEN, AND THE MEDIEVAL TRADITION

Medieval medical theorists followed two interconnected traditions linking the aerial environment to the etiology of disease and human physiology: the Hippocratic tradition acknowledged the presence of internal air but emphasized the power of external terrestrial winds to infect bodies and places; the Galenic tradition emphasized instead the role of internal air, or *pneuma*, in bodily processes of health and illness. Both traditions held that patterns of exercise, sleep, and excretion (or retention), states of emotion, and the environment—specifically the air, the weather, and the winds— were primary factors contributing to humoral imbalance. Galen calls these the "non-naturals," which worked in conjunction with the "naturals" (humors, organs, elements) to prevent or promote the "contra-naturals," namely, disease.[23]

Because the practical Hippocratic tradition was transmitted to the early Middle Ages, albeit in truncated form, through encyclopedic works such as the *Etymologia* of Isidore of Seville and was embedded in patristic and other theological sources from antiquity, the influential nature of terrestrial winds on the body and the noxious powers of infected air became salient features of medieval intellectual culture.[24] The Galenic tradition, with its emphasis on *pneuma*, did not influence medical thinking in the west until the translation movements of the twelfth century and the subsequent introduction of Galenic texts, as well as Arabic commentaries and new compositions, into the burgeoning universities of medieval Europe.[25] In the academic milieu of the thirteenth century, theologians would use ideas shaped by Galenic theories of *pneuma* in conjunction with humoral theory and Aristotelian natural philosophy to argue that internal air might be influenced by the divine or the demonic, and that the toxic *pneuma* of the spiritually corrupt might infect the air and bodies of those surrounding them.

In the Hippocratic tradition, a working knowledge of a patient's regional atmosphere, most importantly the winds and weather, was vital to the practice of medicine. In *Airs, Waters, and Places*, Hippocrates argues that while some might complain that such atmospheric phenomena "belong rather to meteorology," it is through the wind and the weather that the "digestive organs of men undergo a change."[26] Digestion is the central process by which the body regulates the humors; in digestion, the body uses vital heat to transform food into blood, excess blood into sperm and hair, and phlegm into bile.[27] The nature of the winds, then, have a direct bearing on an individual's physical health. In books three through six of *Airs, Waters, and Places*, books one and three of the *Epidemics*, and throughout *On the Sacred Disease*, Hippocrates discourses on the qualities of the cardinal winds, placing particular emphasis on the north wind, which is exceptionally dry, and the south wind, which is extremely moist. Because of their radical dryness and moisture, the north and south winds are held to be the airs most dangerous to human health. According to *Airs*, those dwelling in cities exposed to the dry north winds "between the summer settings and the summer risings of the sun," are cold by nature and predisposed toward dryness.[28] Upon entering the body, the winds of *Septentrio* facilitate the production of black bile, chill the body, rob it of vital heat, and restrict the processes of digestion. Those exposed to the excessive moisture of the southerly "plague" winds, however, are likely to have heads "of a humid and pituitous constitution, and their bellies subject to frequent disorders, owing to the phlegm running down from the head."[29] For Hippocrates, rapid shifts between northerly and southerly winds could wreak havoc on the body, at one moment desiccating it, the next plumping it with fluid;[30] likewise, transitional periods between seasons could disrupt humoral balance, with spring and autumn and the concomitant shift from moisture to dryness and dryness to moisture leaving the body most vulnerable to illness. In the Hippocratic tradition, both shifting winds and seasonal change were the chief causes of acute illness and, in some cases, epidemics.

The air through wind and weather, ever on the move, afflicted the body not only by penetrating it directly but also by corrupting the local environment in which a body dwelled. The Hippocratic assertion that the atmosphere of a given region could become toxic enough to cause bodily harm—a concept rooted in *Airs, Waters, and Places* as well as the *Epidemics*—became a salient feature of ancient discussions of plague. For example, in *De Rerum Natura* (ca. 50 BCE), Lucretius describes the plague of Athens as a "mortal miasma . . . coming from afar / Rising in lands of Aegypt, traversing / Reaches of air and floating fields of foam / At last on all Pandion's folk it swooped."[31] Once settled in Athens, this toxic air affected each differently, "For what to one had given the power to take the vital winds of air into his mouth / And to gaze upward at the vaults of sky / The same to others was their death and doom."[32] In this, Lucretius follows Hippocrates's assertion that plague "comes from the *pneuma* that everyone exhales."[33]

The conceptual linkage between toxic air and plague was transmitted to the medieval world through patristic and hagiographical sources, in which we see the causes of infectious air shift away from the natural, albeit intertwined with the supernatural, to an emphasis on the divine and even the demonic. In his discussion of meteorology in the *Etymologia*, for example, Isidore of Seville echoes Hippocrates in attributing the plague to the southerly winds, "from which corrupt air is born."[34] In his *De Medicina*, however, he argues that while the plague "arises from corrupt air, and by penetrating into the viscera settles there," and that "this disease often springs up from air-borne potencies, *nevertheless it can never come about without the will of Almighty God.*"[35] The belief that superlunary sources caused corrupt and pestilent air remained salient in medieval intellectual culture through the Black Death of 1347, with physicians such as Gentile da Foligno using medical astrology to argue that celestial influences lay behind the plague, while theologians and the broader community of believers continued to cite human sin, divine retribution, and the poisonous darts of God (as first described by Homer) afloat on the foul air as the source of the scourge.[36]

Early authors in the Christian tradition embodied the supernatural agents responsible for the physical corruption of air in the form of serpents or dragons whose poisonous bodies and noxious breath were the simultaneous causes of human disease and death. In an epistle to Bishop John, for example, Pope Gregory the Great (d. 604 CE) complains that a repletion of noxious humors (*noxio humore*) is causing him great suffering, and then turns to a discussion of the plague at Rome, a grave disease caused by putrefied air that has been corrupted by the exhalations of the serpents dwelling in the Tiber River.[37] This same etiology of epidemic disease is echoed in the vita of Saint Sylvester, who descended into a pit near Rome to bind and seal the mouth of a dragon whose stench and breath slew three hundred men a day.[38] The saint who slays the *dracone* and thereby cures a region from a serpent's infectious and venomous breath became a topos in medieval hagiography. In the *Acta Sanctorum*, a collection of Latin and Greek hagiographical sources compiled from manuscripts and incunabula into sixty-seven volumes by the seventeenth-century Bollandists and their descendants, are the tales of Saint Lifard (sixth century CE), who slays a dragon filled with demons (themselves composed of foul air), and Saint Marcel, Bishop of Paris (fourth to fifth century CE), who must kill a dragon that eats corpses and whose breath "infected the air."[39] Likewise, Saint Pavacius, Bishop of Le

Mans (fourth century CE) is faced with a dragon whose "very breath . . . was pestilential"; in order to "rid the country of this plague," Pavacius makes the sign of cross, thereby banishing the beast to the depths of the earth forever.[40] In all of these cases, the toxic breath of the serpent can also be read as an allegory for heresy that, like the plague, spreads quickly and infects/corrupts the souls of the faithful. The link between the toxic breath of an evil creature and its power to poison the surrounding air and kill those who come into contact with it would continue to be employed throughout the later Middle Ages, as theologians working in the milieu of the medieval universities used Hippocratic theories of toxic air, transmitted through both sacred and secular texts, in conjunction with Galenic theories of *pneuma* to explain the spiritual and physical toxicity of evil women.[41]

Galen, unlike Hippocrates, does not emphasize the role of atmospheric winds and regional air conditions in processes of health and disease, but instead focuses on the function of internal air, or *pneuma*, in human physiology. For Hippocrates, bodily air, ever contingent on atmospheric air, is drawn in and exhaled by the veins, the "spiracles of our bodies," which keep the breath moving, a process necessary for physical motion and life.[42] For Galen, however, bodily air is distinct from environmental air and linked closely with cognitive functions such as memory and reason. According to Galen, atmospheric air enters the spongy tissue of the lungs, where it is converted to a more subtle form of air, or *pneuma*. This *pneuma* is then drawn to the left ventricle of the heart, where it comes into contact with superheated blood, which rarefies it into vapor-like vital *pneuma*. From there it is drawn to the brain where it is transformed into two varieties of psychic *pneuma:* sensory and kinetic. In Galenic theory, *pneuma* functions not only as the life force responsible for sense perception and motion but as the fundamental essence of the mortal and material human soul, which he divides, much like Plato, into three elements: the spiritive soul, responsible for courage and altruism, which dwells in the heart; the appetitive soul, the seat of hunger, desire, and irrational emotions, which dwells in the liver; and the rational soul, which resides in the brain. The rational soul not only directs cognitive functions such as imagination, reason, and memory but also theoretically governs the behavior of the spiritive and appetitive souls.

The multivalent nature of *pneuma*, its full integration with the physical body, and its connection to the invisible soul led Galen to believe that spiritual defects, including the inability to control one's emotions and desires, serve as both a symptom and a source of disease which might be discerned through close examination of the body, and vice versa. Galen's *On the Pas-*

sions and Errors of the Soul discusses inordinate emotions as "diseases of the soul. For obstinacy, love of glory, and lust for power are diseases of the soul. Greediness is less harmful than these, but it, too, is, nevertheless, a disease. And what must I say of envy? It is the worst of evils."[43] Also dangerous are unrequited lust and longing, both of which might lead to a disease known as lovesickness, the symptoms of which include insomnia, a deathly pallor, sunken eyes, uneven respiration leading to sighing, as well as a quickened pulse such as one might see in cases of fever; left uncured, lovesickness might cause acute melancholy, with its dark thoughts, aching sorrow, and the physical symptoms associated with excessive black bile. Lovesickness remained a concern in medieval medical treatises, such as Constantine the African's *Viaticum*, a twelfth-century translation of Ibn-Jazzar's *al Hawi*: "The more the soul plunges into these sorts of thoughts, the more its actions and the body's are damaged. The body follows the soul's actions, while the soul follows the body's passions. Whence Galen said, 'The powers of the soul follow the complexions of the body.'"[44]

The reciprocal relationships among the *pneuma*, the human body, and the atmosphere surrounding it were a topic of interest for university medicine, which followed Galen in imagining *pneuma* as finite matter that, albeit imperceptible, behaved according to natural law. So too for academic theology, which viewed *pneuma*, often conflated with the human soul, as ethereal, eternal, and open to the miraculous, and therefore able to supersede natural law. The thirteenth-century Dominican theologian Thomas of Aquinas attempted to reconcile these seemingly dichotomous viewpoints. Following Aristotle, he argues that the immortal soul resides in the tabernacle of the heart and serves as a formal cause for the human body, therefore informing its physical shape; while the soul could not *naturally* exist apart from the body as a separate spiritual entity, it might, however, do so through God's sustenance.[45] For theologians, then, bodily *pneuma* was not merely open to superlunary forces but was in fact a primary conduit for divine communion and communication, a physical manifestation of the breath of God as well as the foundation for the flesh.[46]

For academic theologians steeped in Aristotelian categorical discourse and versant in learned medical theory, the flesh too had its role to play. Following Aristotle, the male body was held to be the paradigm of physical perfection; hot and dry, having firm and powerful flesh, the male physiological system functioned at peak efficiency.[47] Because of its innate heat, the male body digested its food completely into humoral blood, which was then refined into muscle, hair, and sperm, while any excess toxins were purged through sweat. The strength and heat of the properly formed male body aided in the production of a nearly perfect *pneuma*, reflected in the perfect male body and the rational male mind, and protected it from humoral imbalances brought on by environmental factors such as pestilent air. The fortitude and ardent prayer of the pious man, whose heart *pneuma* was infused with the divine, might even protect him from the influence of the cold-bodied demons, themselves composed of cold and foul air, that rode the winds.[48] Women, however, were not so blessed.

WOMEN AND WITCHES

Through the process of Aristotelian categorical inversion, theologians constructed the imaginary female body, in opposition to equally imaginary male perfection, that was completely inefficient: cold and moist, with loose and spongy flesh.[49] Her lack of heat made it impossible to refine fully the blood that resulted from imperfect digestion into hair or muscles, and so she required a receptacle for the storage and periodic disposal of bodily toxins and waste blood.[50] The womb, unique to a woman's anatomy, became the locus of her toxicity, a sign of her physical imperfection and corruption. Since it was cold and moist, a woman's body could not properly heat the air that entered her lungs and convert it to vital *pneuma;* for this reason, her mind was as slow and irrational as her flesh; the categorical woman's cold flesh and underheated *pneuma* left her open to the influences of cold-air demons and the power of pestilent winds. According to medieval theologians, a woman's only hope for protection against such corruption was a chaste life or ardent prayer, the life of a holy woman.

The paradigm of female perfection, the holy woman, is depicted in hagiographical sources as young and virginal, her rosy cheeks a sign of a body and soul warmed by ardent prayer and communion with the divine. One holy woman, Lutgard of Aywières (thirteenth century CE), became so fervent in her devotions that she achieved ecstatic union with Christ, a fiery experience that burned off the dross within her, rarefied her soul, and purified her flesh. In his account of Lutgard's life, the thirteenth-century Dominican theologian Thomas Cantimpré writes that during a moment of spiritual union following upon ardent prayer, a vein ruptured in the Cistercian nun's chest. Lutgard shed such copious amounts of blood that her tunic became soaked. The flow of blood from her side was not only symbolic of her *imitatio Christi* but also of her radical purification, a process that began with the holy communion of Lutgard's spirit with Christ and exploded outward through her physical being, rendering her flesh so miraculously pure that she ceased to menstruate and instead produced substances imbued with the divine, from the oil that dripped from her breasts and fingers to her saliva and breath, all of which were marked by a suave (sweet) odor and all having the power to heal bodies and souls.[51] Infused with divine *pneuma*, Lutgard's miraculous flesh became so light as to float on air like warm and dry elemental fire and, in this levitation, rise closer toward the divine aether that was her mystical home. Lutgard's miraculous experience stands in direct opposition to the Aristotelian dictum that human bodies, composed of a heavy and chaotic mixture of the elements, seek their natural place on the ground. This indicates that her flesh had changed radically in its physical composition. In its purification, the chaotic elements were put to rest, allowing her body to ascend toward the Platonic realms of harmony and, ultimately, to the heavenly court that was her true home. The divine *pneuma* of the theologically constructed holy woman not only made miracles possible and protected her against spiritual corruption but also afforded her physical protection against pestilent air that swirled about sickly places such as hospitals. In another case, the miracles of Saint Elizabeth of Hungary (thirteenth century CE) describe her founding of a hospital for the care of the indigent sick and her work there as a nurse.[52] Her hagiographer writes that day and night she would go down to the hospital to care for the suffering; "risking her health by breathing noxious air, she endured the corruptions of the suffering patients for the love of God, and did not abhor them, but ministered to them, carrying the sick in her own arms so that they could use the facilities, washing their clothes and bedclothes, and tending their suppurating sores."[53] While such pestilent air corrupted by the odors of offal and necrosis might endanger weaker women, causing humoral imbalance and even death, saintly women were physically and spiritually fortified by the divine *pneuma* that flowed throughout their miraculous beings.

The warmth of prayer and the fire of the Holy Spirit had the power to heal women of their fundamental coldness, the source of their physical, mental, and spiritual corruption, and to protect them from harm. While devout women in controlled environments, such as Lutgard of Aywières, and saintly women graced by God, such as Elizabeth of Hungary, might attain the purity of the divine, most women left to their own devices were likely to fall victim to their cold, moist natures and the temptations of the flesh. Just as theologians constructed the imaginary corrupt female body in opposition to that of the imaginary perfect male, so too did they use Aristotelian categorical inversion to construct the evil and toxic woman in opposition to that of the holy and chaste virgin-saint.[54] In creating the paradigmatic evil woman, thirteenth- and fourteenth-century scholars nonsystemati-cally applied Aristotelian natural philosophy, humoral theory, and anatom-ical knowledge to prove the primarily theological assertion that the female soul is attracted to evil and that the female body was so toxic that it might poison those around it.[55] One such scholar was Pseudo-Albertus Magnus, the thirteenth-century author of *De Secretis Mulierum*, a treatise on the purported secrets of women. Pseudo-Albertus Magnus premises much of his discussion on the weak, irrational, and wanton nature of the female soul, which is open to the subtle lies of demons; this essentially corrupt fe-male soul forms a body that is cold, moist, and suffers from a chronic im-balance of humors. The physical locus of the corrupt woman, of course, is her uterus and the poisonous by-products of menstruation that it contains and routinely expels.[56] While medical authors such as Hippocrates, Galen, and Avicenna argue that menstrual blood was natural and harmless, theo-logians such as Isidore of Seville had long claimed that it was toxic enough to cloud mirrors, rust metal, and dissolve glue.[57] Pseudo-Albertus agrees with Isidore and misquotes Hippocrates's *On the Nature of Man* in support of his argument that men must avoid sexual congress with menstruating women, "because from this foulness the air is corrupted, and the insides of a man are brought to disorder."[58] For Pseudo-Albertus, direct contact with menstrual blood would expose the perfectly balanced male *pneuma* to toxic vapors which might then poison his internal air, corrupt his soul, disrupt his humoral balance and, through this, his physical well-being.

Even more dangerous than a woman's regular flow is menstrual fluid that remains in her uterus against nature; if not voided on a monthly ba-sis, the menses begin to fester and produce noxious fumes that rise along the inner passages of the body, called *phlebes*, thereby reaching the breasts, mouth, nose, eyes, and brain.[59] Younger women are prone to this condition, but it is far more common in older women, whose "natural heat is so defi-cient that the menses collected in them cannot be expelled."[60] According to Pseudo-Albertus, "the venomous humors" from such a woman's fetid men-strual fluid have the power to "infect the air by her breath"; this "infected air" might then travel outward to corrupt the airways of men, causing cough and other respiratory ailments, and to poison infants who come into contact with it. Old women, Pseudo-Albertus argues, should not be allowed near vulnerable infants and children because "they poison the eyes of children lying in their cradles."[61] The reason, one of his commentators elaborates, is that an old woman's "menses are venomous" and therefore "continually borne to the eyes. Because of the porosity of the eyes, they infect the air, which reaches the child, for he is easily infected because of his tenderness."[62] Medieval medical theory supported Pseudo-Albertus's academic version of the evil eye. According to medical theory, psychic *pneuma* left the brain through the optic nerve, came into contact with physical objects, and returned to the brain to form an image.[63] In toxic women, noxious uterine vapors were intermingled with the *pneuma* that served as the matrix for vision; because the process of vision was an intimate one, involving physical contact between the viewer and the object viewed, spiritual and physical contagion by poisonous *pneuma* was a terrifying and very real possibility.

For theologians like Thomas Aquinas, the corrupt air exuded by toxic women might not only contain foul bodily vapors, the result of natural processes unique to female physiology, but also filthy spirits, or demons. Themselves composed of cold coagulated noxious air, demons and their master, Satan, could be discerned by their icy flesh and foul breath. Aquinas argues that women, weak and wanton, would offer themselves sexually to demonic beings, which would tempt them in the guise of men. In this form, the demon might penetrate a sinful woman's loose and cavernous body, disfigure her *pneuma*, poison her soul, and render her even more toxic than her natural composition suggested.[64] A demon-infested woman's foul *pneuma*, which shaped the material form of her body, was written upon her icy, green, and scabrous flesh and tangled in her serpentine hair. Together, the authoritative demonology of Aquinas in conjunction with treatises on women, such as the *De Secretis Mulierum* of Pseudo-Albertus, provided a foundation for medico-theological theories of witchcraft. In Johannes Nider's *Formicarius* of 1437, for example, we find the first *explicit* connections between corrupt female physiology, female wantonness and irrationality, and formal collusion with demons and the devil.[65] In the clerical imaginations of men like Nider, the evil woman—no longer merely the victim of temptation—is ever surrounded by hovering demons, fetid and foul, awaiting her command.[66] Through her demonic pact, the witch—who is otherwise powerless save for her toxicity—becomes the master of natural forces such as storms and wind, which are themselves laden with invisible evil entities. Nider's ideas are reified in Kramer and Sprenger's *Malleus Maleficarum* (1486), an inquisitorial manual that came to define witchcraft for a clerical and lay audience, as well as in Martin Castañega's *Treatise on Witchcraft* (1529) and in Reginald Scot's *The Discoverie of Witchcraft* (1584), all of which—in their own way—proliferated the connections between the imaginary biology of the witch, her pestilent breath, and her command of the foul and filthy air, upon which she might even ride.

Double, double, toil and trouble. Like the ouroboros, we have come full circle, from plague-breathing dragon to pestilent witch, each of whose serpentine and foul bodies were so toxic that they might poison those who came into contact with them. Like Gregory the Great's serpents poisoning the Tiber, the monstrous body of the witch was a foul pit of noxious humors and vapors whose very glance, breath and stench had the power to breed pestilence. The frigid witch, conjured from the quill pens of thirteenth- and fourteenth-century clerics, still rides the winds of our imaginations, from the Weird Sisters of *Macbeth*, with their demonic foresight, command of the wind, and poisonous breath, to the White Witch of C. S. Lewis's *Lion, the Witch, and the Wardrobe* to Serafina Pekkala, Phillip Pullman's witch queen who rides the wind and controls the weather.[67] The strange powers and toxic natures of these imaginary creatures, however seemingly harmless and playful to us today, have a dark and bloody history. Rooted in Aristotelian natural philosophy and academic medical theory as interpreted and applied by late medieval theologians, the witch is a monster stitched together by male academics in their quest to rationalize long-held beliefs about the nature of pestilence, poisonous flesh, the danger of women, and the terrifying potential for evil to be hidden in that invisible, vital, amorphous, and ubiquitous element, *air*.

NOTES

1. *Macbeth*, 1.1.11–12. References are to act, scene, and line.2. *Macbeth*, 1.3.25.

3. Holinshed, *Chronicles*, V.

4. James I, *Daemonologie*, chapter 5.

5. Before the thirteenth century, witches were cunning women who were heretical in their beliefs and who fell victim to the illusions of demons, but neither witches nor demons were believed to be particularly powerful. As thirteenth- and fourteenth-century scholastic theologians used Aristotelian contrariety, natural philosophy, and learned medicine to define ultimate good and basest evil, they created a binary model that would ultimately be used to create the maleficent, melancholic, and Satanic witch in the early fifteenth century. See Clark, *Thinking with Demons*, and Bailey, "Medieval Concept of the Witches' Sabbath."

6. Grant, *Nature of Natural Philosophy*, 29–34; Leff, *Paris and Oxford Universities*.

7. Aristotle, *Metaphysics*, 6.7, available online through the MIT Internet Classics Archives at http://classics.mit.edu/Aristotle/metaphysics.html.

8. Aristotle, *Metaphysics*, 1.3, 1.7; Aristotle, *On Generation and Corruption*,

1.1 and 2.1, available online through the MIT Internet Classics Archives at http://classics.mit.edu/Aristotle/gener_corr.mb.txt.

9. Aristotle, *Meteorology*, 1.2, available online through the MIT Internet Classics Archives at http://classics.mit.edu/Aristotle/meteorology.html.

10. Wildberg, *John Philoponus' Criticism*, 12.

11. On the complex names for the variety of winds and airs see Lloyd, "Pneuma," S136.

12. Aristotle, *Meteorology*, 1.3, 1.9.

13. Aristotle, *De Caelo*.

14. Aristotle, *Meteorology*, 1.3, 2.4.

15. Obrist, "Wind Diagrams," 36.

16. Isidore of Seville, *De Rerum Natura*, 8th century, Munich, Bayerische Staatsbibliothek, MS 16128, fol. 16r.

17. On Boreas, see Hesiod, *Works and Days*, lines 474–563.

18. Obrist, "Wind Diagrams." While angels might descend to earth from the ethereal realms beyond the moon, they did not swirl about in the air, nor did they linger for long; only demons, trapped in this world below the moon, dominated the air in the earthly realm.

19. Isidore of Seville, *Differentiarum II*, 77–78.

20. On thirteenth-century academics and the Aristotelian qualities of the moon in its different phases, see Lemay, *Women's Secrets*, 28; see also Bartholomew the Anglican, *De Proprietatibus Rerum*.

21. The *mundus-annus-homo* diagram is now the emblem of the Warburg Institute, London. "The Emblem of the Warburg Institute," Warburg Institute, University of London, School of Advanced Study, accessed Sept. 18, 2013, http://warburg.sas.ac.uk/home/aboutthewarburginstitute/emblem/.

22. On the *Viaticum* and the *Liber Pantegni*, see Burnett and Jacquart, *Constantine the African*. On the works of Rhazes and Avicenna, see Savage-Smith and Pormann, *Medieval Islamic Medicine*. On the *Lilium Medicinae*, see DeMaitre, *Doctor Bernard de Gordon*.

23. Ballester and Arrizabalaga, *Galen and Galenism*.

24. Lindberg, *Beginnings of Western Science*, 321–22.

25. Ballester and Arrizabalaga, *Galen and Galenism*; Bono, "Medical Spirits."

26. Hippocrates, *Airs, Waters, and Places*, part 2.

27. See Langholf, *Medical Theories*, 84.

28. Hippocrates, *Airs, Waters, and Places*, bk. 4

29. Ibid., bk. 3

30. Hippocrates, *On the Sacred Disease*.

31. Lucretius, *De Rerum Natura*, bk. 6.4.

32. Ibid.

33. Hippocrates, *On Breaths* and *On the Nature of Man*, ch. 9; Lloyd, "Pneuma," S138.

34. Isidore of Seville, *Etymologies*, bk. 13.11.6.

35. Emphasis mine. Isidore of Seville, *Medical Writings*, 32.

36. Horrox, *Black Death*; Homer, *Iliad*; Jolly, *Popular Religion*.

37. Gregory the Great, letters to Bishop Mariano of Arabia and Rusticanus Patricie, Biblioteca Nacional de España, Madrid, MS 11544, folio 13v.

38. Jacob de Voragine, *Legenda Aurea*.

39. On Saint Lifard, see *Acta Sanctorum*; on Saint Marcel, see Gregory of Tours, *Historiae Francorum*, bk. IX.

40. Brewer, *Dictionary of Miracles*, 116.

41. Hippocrates, *On Breaths* and *On the Nature of Man*, ch. 9; Lloyd, "Pneuma," S138.

42. Hippocrates, *On the Sacred Disease* and *On Breaths*.

43. Galen, *Errors and Passions*, 53.

44. Wack, "*Liber de heros morbo* of Johannes Afflacius," 329.

45. Thomas of Aquinas, *Summa Contra Gentiles*, 68. See also Thomas of Aquinas, *Summa Theologica*, trans. Fathers of the English Dominican Province, 1920, question 75, http://www.newadvent.org/summa/.

46. Thomas of Aquinas, question 91, article 4, reply objection 3.

47. Aristotle, *Politics*, 1.5, available online through the MIT Internet Classics Archives at http://classics.mit.edu/Aristotle/politics.html; Aristotle, *On the Generation of Animals*.

48. On the tightly packed and sealed nature of male bodies, see Caciola, "Mystics," 290.

49. See Hacking, "Aristotelian Categories" (postmodernist); Cadden, *Meanings of Sex Difference* (fuller treatment); and Clark, *Thinking with Demons* (Aristotelian contrariety).

50. Dean-Jones, *Women's Bodies*.

51. Thomas of Cantimpré, *Collected Saints' Lives*.

52. Jacob de Voragine, *Legenda Aurea*, 1159.

53. Ibid., 1162–63.

54. On witches as inverted saints, see Kieckhefer, "Holy and the Unholy."

55. Lemay, *Women's Secrets*.

56. Dean-Jones, *Women's Bodies*.

57. Isidore of Seville, *Medical Writings*, 48.

58. Lemay, *Women's Secrets*, 88.

59. Dean-Jones, *Women's Bodies*.

60. Lemay, *Women's Secrets*, 131.

61. Ibid.

62. Ibid.

63. Smith, "Alhacen's Theory."

64. On the spongy female body as a locus for demons, see Caciola, "Mystics."

65. Bailey, *Battling Demons*.

66. Bailey, "From Sorcery to Witchcraft," 978.

67. Pullman, *His Dark Materials*.

REFERENCES

Acta Sanctorum: quotquot toto urbe coluntur, "De SS. Lifardo et Urbico," June 3, Tome 1, 291–301. Paris: V. Palmé, 1863–1940.

Aristotle. *Metaphysics*. Translated by W. D. Ross. Oxford: Clarendon Press, 1924.

———. *Meteorology*. Translated by E. W. Webster. Oxford: Oxford University Press, 1963 (reprint).

———. *On Generation and Corruption*. Translated by H. H. Joachim. Oxford: Clarendon Press, 1922.

———. *On the Generation of Animals*. Translated by A. L. Peck. Cambridge, MA: Harvard University Press, 1953.

———. *Politics*. Translated by Benjamin Jowett. Oxford: Clarendon Press, 1885.

Bailey, Michael. "The Medieval Concept of the Witches' Sabbath." *Exemplaria* 8, no. 2 (1996): 419–39.

———. "From Sorcery to Witchcraft: Clerical Conceptions of Magic in the Later Middle Ages." *Speculum* 76, no. 4 (2001): 960–90.

———. *Battling Demons: Witchcraft, Heresy, and Reform in the Late Middle Ages*. University Park: Pennsylvania State University Press, 2003.

Ballester, Luis García, and John Arrizabalaga, eds. *Galen and Galenism: Theory and Medical Practice from Antiquity to the European Renaissance*. Burlington, VT: Ashgate Press, 2002.

Bartholomew the Anglican. *De Proprietatibus Rerum*. Frankfurt-am-Main: Minerva Press, 1964.

Bono, James J. "Medical Spirits and the Medieval Language of Life." *Traditio* 40 (1984): 91–130.

Brewer, E. Cobham. *A Dictionary of Miracles: Imitative, Realistic, and Dogmatic*. Philadelphia: Lippincott, 1894.

Burnett, Charles, and Danielle Jacquart, eds. *Constantine the African and Ali ibn al-Abbas al-Magusi: The Pantegni and Related Texts*. Studies in Ancient Medicine 10. Leiden: E. J. Brill, 1994.

Butler, Alban. *Lives of the Fathers, Martyrs, and Other Principal Saints.* London, 1815.

Caciola, Nancy. "Mystics, Demoniacs, and the Physiology of Demon Possession," *Comparative Studies in Society and History* 42, no. 2 (2000): 268–306.

Cadden, Joan. *Meanings of Sex Difference in the Middle Ages: Medicine, Science, and Culture.* New York: Cambridge University Press, 1993.

Clark, Stuart. *Thinking with Demons: The Idea of Witchcraft in Early Modern Europe.* Oxford: Oxford University Press, 1999.

Dean-Jones, Lesley. *Women's Bodies in Classical Greek Science.* Oxford: Clarendon Press, 1996.

DeMaitre, Luke. *Doctor Bernard de Gordon: Professor and Practitioner.* Toronto: Pontifical Institute of Mediaeval Studies, 1980.

Galen. *On the Errors and Passions of the Soul.* Translated by Paul Harkins. Columbus: Ohio State University Press, 1963.

Grant, Edward. *The Nature of Natural Philosophy in the Late Middle Ages.* Washington, DC: Catholic University of America Press, 2010.

Gregory of Tours. *Historiae Francorum libri decem.* Paris: G. Morellium, 1561.

Hacking, Ian. "Aristotelian Categories and Cognitive Domains." *Synthese* 126, no. 3 (2001): 473–515.

Hesiod. *Works and Days.* Translated by David Tandy and Walter C. Neale. Berkeley: University of California Press, 1996.

Hippocrates. *On Breaths.* In *Hippocrates, Volume II: Ancient Medicine.* Translated by W. H. S. Jones. Loeb Classical Library 148. Cambridge, MA: Harvard University Press, 1923.

———. *On the Sacred Disease.* In *Hippocrates, Volume II: Ancient Medicine.* Translated by W. H. S. Jones. Loeb Classical Library 148. Cambridge, MA: Harvard University Press, 1923.

———. *Airs, Waters, and Places.* In *Hippocrates, Volume I: Ancient Medicine.* Translated by W. H. S. Jones. Loeb Classical Library 147. Cambridge, MA: Harvard University Press, 1923.

———. *On the Nature of Man.* In *Hippocrates, Volume IV: Nature of Man.* Translated by W. H. S. Jones. Loeb Classical Library 150. Cambridge, MA: Harvard University Press, 1931.

Holinshed, Raphaell. *Chronicles: England, Scotland, Ireland.* Vol. 5, *Scotland.* New York: Routledge, 1965.

Homer. *Iliad.* Translated by Anthony Verity and Barbara Graziosi. New York: Oxford University Press, 2012.

Horrox, Rosemary. *The Black Death.* Manchester: Manchester University Press, 1994.

Isidore of Seville. *Differentiarum II*. In *Patrologiae Latinae* 83, edited by J. P. Migne. Paris: Apud Gamier Fratres, 1895.

———. "The Medical Writings. An English Translation with an Introduction and Commentary." Translated and with an introduction and commentary by William D. Sharpe. *Transactions of the American Philosophical Society* 54, no. 2 (1964): 1–75.

———. *Etymologies*, Translated by Stephen A. Barney. Cambridge: Cambridge University Press, 2006.

Jacob de Voragine. *Legenda Aurea*. Edited by Giovanni Paolo Maggioni. Florence: SISMEL, 1998.

James I. *Daemonologie in forme of a dialogue*. Edinburgh: Robert Walde-Grave, 1597.

Jolly, Karen Louise. *Popular Religion in Late Saxon England: Elf Charms in Context*. Chapel Hill: University of North Carolina Press, 1996.

Kieckhefer, Richard. "The Holy and the Unholy: Sainthood, Witchcraft, and Magic in Late Medieval Europe." *Journal of Medieval and Renaissance Studies* 24 (1994): 355–85.

Langholf, Volker. *Medical Theories in Hippocrates: Early Texts and the "Epidemics."* Berlin: Walter de Gruyter, 1990.

Leff, Gordon. *Paris and Oxford Universities in the Thirteenth and Fourteenth Centuries: An Institutional and Intellectual History*. New York: Wiley, 1968.

Lemay, Helen Rodnite. *Women's Secrets: A Translation of Pseudo-Albertus Magnus's De Secretis Mulierum with Commentaries*. Albany: State University of New York Press, 1992.

Lindberg, David. *The Beginnings of Western Science: The European Scientific Tradition in Philosophical, Religious, and Institutional Context, Prehistory to AD 1450*. Chicago: University of Chicago Press, 2007.

Lloyd, Geoffrey. "Pneuma between Body and Soul." *Journal for the Royal Anthropological Institute* 13, no. S1 (2007): S135–46.

Lucretius. *De Rerum Natura*. Translated by W. H. D. Rouse. Loeb Classical Library 181. Cambridge, MA: Harvard University Press, 1924.

Maggionne, Giovanni Paolo, ed. *The Miracles of Saint Elizabeth of Hungary in the Legenda Aurea of Jacob de Voragine*. Florence: SISMEL, 1998.

Obrist, Barbara. "Wind Diagrams and Medieval Cosmology." *Speculum* 72, no. 1 (1997): 33–84.

Pullman, Phillip. *His Dark Materials Omnibus*. New York: Alfred A. Knopf, 2007.

Savage-Smith, Emilie, and Peter Pormann. *Medieval Islamic Medicine*. Edinburgh: Edinburgh University Press, 2007.

Shakespeare, William. *Macbeth*. New York: Oxford University Press, 2008.

Smith, Mark A. "Alhacen's Theory of Visual Perception: A Critical Edition, with

English Translation and Commentary, of the First Three Books of Alhacen's *De aspectibus,* the Medieval Latin Version of Ibn al-Haytham's *Kitāb al-Manāẓir.*" *Transactions of the American Philosophical Society* 91, no. 4 (2001): i–337.

Thomas of Aquinas, *Summa Contra Gentiles: Book Two.* Translated by James F. Anderson. Notre Dame: Notre Dame University Press, 1992.

Thomas of Cantimpré. *The Collected Saints' Lives: Abbot John of Cantimpré, Christina the Astonishing, Margaret of Ypres, and Lutgard of Aywières.* Translated by Barbara Newman and Margot King. Brepols: Turnhout, 2008.

Wack, Mary Frances. "The *Liber de heros morbo* of Johannes Afflacius and Its Implications for Medieval Love Conventions." *Speculum* 62, no. 2 (1987): 324–44.

Wildberg, Christian. *John Philoponus' Criticism of Aristotle's Theory of Aether.* Berlin: De Gruyter, 1988.

2

SURGEON REGINALD ORTON AND THE PATHOLOGY OF DEADLY AIR
The Contest for Context in Environmental Health

Christopher Hamlin

THE WORKS ON cholera of the Anglo-Indian military surgeon Reginald Orton (1790–1835) and of like-minded successors over the remainder of the nineteenth century have much to tell us about why issues involving air and disease have remained troublesome in our own era, both as matters for scientific study and as issues for public response. The goal of this chapter is to explore the problematic status of the atmosphere as a cause of ill health in Western medical heritage before the contemporary era of environmental awareness. From the reception of Orton's work, and that of others engaged in analogous projects, we can learn much about why issues involving atmosphere/climate and disease remain marginal in our own era, equally as research questions and as areas of policymaking.[1] Orton worked at the intersection of ancient medical climatology and modern biomedical reductionism.[2] His research might be seen as a pioneering exemplar of a holistic and interdisciplinary paradigm of infectious disease ecology, but instead represents a genre that has been almost completely forgotten.[3] Accordingly it is worth exploring its marginalization, both within medical history and more broadly with regard to the ways modern publics are invited to consider the intersection of environment and health.

I shall suggest that the reasons for the continued troublesomeness in identifying atmospheric/climatic determinants of disease are both epistemic and contingently historical. With regard to the former, Orton and his contemporaries and colleagues were multicausalists in an era long before mathematical means were available to distinguish the significance and interconnectedness of multiple causes. With regard to the latter, Orton's story is one of the geography of professional authority. In an age in which scientific medicine was being consolidated in European hospitals and laboratories, explorers of exotic colonial airs and climates were marginalized. Condemned by context in their own age, most remain so in modern histories of medicine in which the colonial remains peripheral and derivative, in which acute attacks are more interesting than chronic exposure, and in which climates/atmospheres are presumed to be innocuous, or, at most, incidental to disease, which is paradigmatically understood as encounters with microbe agents.[4] I shall suggest here not only the importance of resisting the temptation to read colonial medicine exclusively in terms of the European mainline but also the importance of recognizing ways of medical knowing that remained prominent and powerful in colonial settings but were never well integrated into European medicine. Studies of atmospheric and climatic determinants of disease exemplify such approaches.

Today's popular press has seemingly reversed assumptions of climatic/atmospheric innocuousness, but has done so incoherently. A series of recent headlines describes the formidable impact of air pollution on human health in a multiplicity of ways: air pollution "kills"; it "causes," is "responsible for," "is linked to," or "contributes to" deaths.[5] The American Lung Association asserts that "breathing polluted air can seriously harm your health and even shorten your life."[6] And its 2011 "State of the Air" report cautions that "over 154 million people [in the United States]—just over one half the nation—suffer pollution levels that are too often dangerous to breathe."[7] The report does not identify a preferred short-term alternative.

Such claims are typically vague and unquantified. Is "harmful to health" the same as "causing disease"? Is shortening life the same as causing death? How much damage to health and how many additional deaths are there? The World Health Organization recently estimated that air pollution causes two million "premature deaths" globally per year.[8] But here, such gross quantification, so often looked to as the basis for public policy, is complicated by the multiplicity of methodologies employed. Some approaches are prospective and toxicological: they measure exposure to particular pollutants. Others are retrospective and epidemiological: they compare mortality and morbidity against a variety of measures of atmospheric and environmental quality. Ideally, these approaches should be complementary, but making them so in fact requires juggling a wide range of complicating issues. It also requires confronting philosophical issues of causation—do we look at causes synergistically or individually? How do we assess blame and responsibility when causes are multiple and in most cases long-term and indirect?

In part, the difficulties encountered in debating and responding to such issues reflect the ways the history of the medical sciences has been written and understood. The picture of the medical past implicit in the headlines excludes, marginalizes, or misrepresents studies of the pathological effects of airs and climates. In its worst incarnations, such a view fetishizes a history in which medical progress is equated with:

> the sharp distinguishing of one disease from others (famously due to recognition of the distinct lesion each makes by the anatomists of nineteenth-century Paris)

> an agenda of simplifying cause-effect relationships by emphasizing acute exposure to single factors over chronic exposure to multiple factors (evident in the attempted attribution of each disease to a single microbe- or toxin-agent of each disease)

> institutions of accountability and prophylaxis oriented toward the single intervention, which, by blocking operation of the cause of a particular disease, will prevent the effect

Airs defy these conventions. Their effects may include, but cannot be reduced to, conditions marked by a single distinctive lesion of structure or function. Moreover, they are likely to be cumulative over a long term and will be coupled with the effects of other constitutional and chronic causes of ill health. Finally, air is the master commons: often its quality can be attributed to no single party or agency, but rather represents integrative characteristics of natural circumstances, modes of industry, the spatial organization of the built environment, sources and uses of energy, and accumulated personal choices. A comparison with water pollution will drive home the point. The effects of bad water can much more easily be traced to a particular microbe or microbes or pollutant, producing a particular disturbance to structure and function, which enters water sources used by humans in a finite number of ways, even perhaps from a single source.[9]

A curious anomaly of medical history is the representation of environmental health concerns simultaneously as very recent—as predicated on developments in the environmental sciences; and as very ancient, exotic, and outdated—an exemplar of a prescientific age. Often explorations of atmospheric determinants of health do not even make it into the catalogue of great medical errors. When they do, their usual fate is to be quickly dismissed; most historians treat them as a reminder of the great wasteland of futile empiricism or suggestive correlation.[10]

This anomaly reflects the programmatic uses of medical history. Long a subsidiary area of medicine itself, medical history was no mere recording of the past, but was a framing and justifying of the present and an agenda for the future. As such, retrospection and tunnel history loom large: they are important as means to articulate and defend a chosen agenda. Traditional medical historiography caricatures events and investigations as wrong and wrongheaded if they seem not to belong to the mainline of medical advance, or alternatively as initial or positive steps (if incomplete contributions) if they fit the dominant discourse. Because the history of medicine is part of a critical public dialogue about health policies, such double standards may be unavoidable, even necessary. My concern here is not that medical history be decoupled from contemporary need but that the tales historians have been telling are of little service in helping us take seriously atmospheric/climatic impacts on health. Hence there is value in recontextualizing past investigations in such a way that they frame the environmental health agendas of the present.

In the nineteenth century one sees the abandonment of an older tradition in environmental medicine, ultimately to be superseded by a mode of laboratory-based microbiology that was often slow to recognize the larger agenda of disease ecology. That era was one of consolidation of medical authority into the metropole, first into large hospitals where clinical research could be carried out; later into the laboratory. It was also the great era of colonization. In terms of Bruno Latour's idiom, the pasteurization was not only of France but of the rest of the world, and it was a "Kochization" as well.[11]

Order took hold by denying or exoticizing the disorder represented in the vicissitudes of atmospheres and climates, often in faraway places. As a detachable segment of the Hippocratic heritage, the famous text *Airs, Waters, and Places* was readily exoticized, in line with Edward Said's "orientalism."[12] The ancients had paid attention to places and airs; so still did South Asian and Chinese classical medical traditions (in which many diseases are matters of winds).[13] By the end of the nineteenth century the discovery of microbial agents suggested that the physical and chemical qualities of the environment itself had no significant potency for health or for illnesses. Rather, this was largely the result of transmission of incidental and organic particulate entities. This reduction of pathology to pathogens, detectable only by experts, had the effect of representing the atmospheric and climatic factors pointed to equally by ancient Greeks and by modern Asians as charming, prescientific approximations. The substitution of knowledge of microbes and their vectors for vague fears of deadly places or airs would become an icon of the arrival of medical modernity. It was also an important justification both for institutions of biomedical science and for colonial rule.

A good illustration is the trajectory of malaria, which generally stands in as the poster child of wrongheaded climatic explanations. To its most influential nineteenth-century articulator, John Macculloch, the existence of a poisonous state of atmosphere (a mal-aria) had been an axiom; such a poison must exist to cause otherwise inexplicable fevers of place; equally it must cause some disease response in any person exposed to it. Convinced of the omnipotence of this aerial poison, Macculloch devoted much effort to discovering its subacute manifestations since not everyone in malarial regions developed the full-blown fever.[14] But by end of century that poisonous air had been triumphantly translated into a mere plasmodium-bearing mosquito. The taming of malaria—in terms of scientific knowledge if not effective prophylaxis—was assumed to exonerate air (and climate), which was innocuous if *Anopheles*-free. Yet, during the same years, a common and deadly condition of environmental exposure often confused with malaria, and known as "thermic fever" (to be considered below) was being relegated to the medical margins. The term disappeared; replacement of thermic "fever" by "heat stroke" or "sunstroke" was a recognition that the pathologic state was not a proper disease, but an incidental interaction between body and air, more likely a matter of exposure than scientific medicine. The atmosphere—here primarily temperature and humidity—is safe; just don't be out in it.

A similar asymmetry of assessment occurred with regard to methods of biomedical science, the dimension I will explore most closely here. The authority for the doctrinal triumph came from the methods of inquiry adopted in hospitals, statistical bureaus, and laboratories, which, it was assumed, must sooner or later bring a comprehensive understanding of the nature of disease. Again, the icons are familiar: the "numerical" method developed in the Paris hospitals constituted righteous repudiation of speculation and theorizing and replaced them with sound induction via the gathering and ordering of facts. The epidemiology of John Snow, the famous mid-century *Introduction to the Study of Experimental Medicine* of the physiologist Claude Bernard, and the postulates of Robert Koch all insisted on the demonstration of a determinate relation between cause and effect. But often overlooked within such narratives are the conflicts among the espousers of rival sorts of rigor and the limitations in adequacy and applicability in their particular methods. More importantly, for most of the nineteenth century and for two centuries before, the same motivations and methods were being applied to the health effects of atmospheres and climates, rendering them part of scientific medicine rather than part of what it replaced.

The clearest example of a double standard comes with regard to empiricism. While the empiricism of the Paris clinicians has been celebrated, that of neo-Hippocratic investigators in the seventeenth through nineteenth centuries has been seen as naïve, a curious manifestation of materialist ideology in the case of Volney or Montesquieu; alternatively as an illustration of new Humboldtian forms of scientific organization, which appear the more remarkable when one contrasts the vastness of the data-gathering enterprise with the paucity of product—knowledge deemed useful for biomedical advance.[15] Delight in being able to measure atmospheric variables had outstripped the ability to analyze data. Correlations of atmospheric variables with mortality or morbidity of certain types did not obviously translate into understandings of pathological mechanisms or therapeutic/prophylactic policies. But nor, in many cases, did the correlations of the Paris clinicians. The knowledge of the investigators of health, place, and atmosphere might at least be useful to travelers, and for the siting of settlements or military bases.

What warrants my focus on the forgotten Orton, surgeon to a regiment in the army of the East India Company in the 1810s and 1820s and author of one of the first grand treatises on cholera, is that he is much more than a fact gatherer and instead comes close to representing the full complement of methodological approaches that figure in medical histories as the explanatory agenda of reductionist biomedical science. Thus, he is both a good illustration of the double standard in historical assessment and a good subject for an exploration of what explanatory expectations we should have in linking climate to disease.

During the nineteenth century conventions of adequate explanation of disease were reconceived in terms of scientific achievement. Medical science did become able, at least in certain diseases and certain ways, to link internal (anatomy, physiology) with external (etiology); that achievement became normative. Henceforth, explanation would involve answering at least four discrete questions: First, can one actually measure independently the entity suspected of harming health? Second, can one posit a mechanism for how it affects the body? Third, are there means for its incorporation into an experimental system of pathophysiology that will allow hypotheses about the mechanism to be confirmed and explored? Fourth, are there adequate means of epidemiological correlation, for field testing of its actual effect on human populations?[16] Orton tried his hand at all of these.

THE CONTEXT: ORTON AND ETIOLOGICAL EXPLANATION

It is important to view Orton against the backdrop of the history of the philosophy of medicine. The period in which he was active was arguably the last in which comprehensive questions of the philosophy of medical inquiry were seriously discussed. Thereafter epistemic issues would be addressed as questions of experimental design/inferential practice within subdisciplinary contexts—for example, pathology, bacteriology, toxicology, and social epidemiology, but also immunology and genetics. The philosophical discussions that peaked in the early nineteenth century had begun in the seventeenth, with the emergence of incompatible pathological theories. Some emphasized mechanical wear and tear (e.g., Bellini, Pitcairne, Cheyne, Hoffmann), others the specificity of fermentable chemical entities (e.g., van Helmont, Willis), and still others the operation of vitalized structures and fluids (e.g., Stahl, Haller, Bordeu, Cullen).

From the later 1660s on, one common response, associated with Thomas Sydenham, was to disengage with speculative pathology; instead to simply describe diseases and to correlate diseases and deaths with time, place, and external conditions, including states of atmosphere. From Sydenham's standpoint, the Baconianism was successful. From studies of the diseases of London in the 1660s and 1670s he inferred the existence of distinct atmospheric "constitutions," each lasting a few years, which affected the character of endemic and epidemic disease. Writing to defend clinical observation against theoretical disease concepts and pathological explanations relying on ungrounded abstractions, Sydenham defied any attempts to reduce these atmospheric constitutions to their components: "This is how it is. There are different constitutions in different years. They originate neither in their heat nor their cold, their wet nor their drought; but they depend upon certain hidden and inexplicable changes within the bowels of the earth." A few pages later: "Still more impossible . . . would it be for me to describe specific forms of epidemics as arising out of specific changes of the atmosphere; easy as such a proceeding may appear to those who can theorise about fevers. . . . The speculations of reason, in and of themselves of paramount greatness and transcendency, is what we do *not* require."[17] It was safer not to pretend to understand. Though positing a geophysical/atmospheric explanation of disease, he would ignore all the elements I have identified above: the isolation of an atmospheric factor, the question of mechanism, the possibility of experiment, the labor of epidemiological correlation.

Sydenham's manifesto to forsake speculation for clinical description would be widely revered as the empirical deck-cleaning that medical science needed. It was both an early example of and inspiration for the eighteenth-century Hippocratic revival.[18] Not only the famous *Airs, Waters, and Places*, but the *Epidemics* reflected concern for winds, heat, and humidity.

Sydenham's patent lack of interest about what in the skies might be manifesting in changing patterns of disease reflects in part the fact that there was not yet much to measure. He wrote roughly a century before the atmosphere would be understood merely as a mix of gases with various particulates floating about, and with peculiar electrical properties as well. His appeal to atmospheric causes reflected a longstanding practice of explaining plagues. General changes in disease required a general cause: only the atmosphere was sufficiently general.

The expansion during the eighteenth century of the range of measurable atmospheric variables (chiefly the gases disclosed by the new pneumatic chemistry) brought hope that the secret atmospheric causes of disease might yield to vast data-gathering campaigns.[19] The most important of these, prominent from the early 1770s until 1830 or so, were those of urban eudiometric investigation. "Eudiometry" comprised a number of means for assessing air quality instrumentally, based on a variety of working principles. Joseph Priestley's approach, based on mixing an air sample with nitrous gas, measured respirability. Others, following Alessandro Volta, were testing for flammable mephitic gases, particularly those given off by putrefaction. Volta eudiometers were reasonably portable and allowed one to measure atmospheric danger from place to place and in a single place over time, and to correlate one's findings with other atmospheric variables and with disease incidence.[20] By the 1830s there was a considerable body of research, much of it done by Paris-trained toxicologist-physiologists François Magendie and Manuel Orfila, in which laboratory animals were exposed to specific gases—chiefly mephitic, phlogisticated, or azotic airs that might arise in crowded towns. Generally, however, this work would be seen as a dead end. While the gases were deadly in sufficient concentration, the dying animals did not adequately replicate symptoms of particular diseases. Toxicology would remain off the mainstream of medicine; even in cases where effects of acute exposure were dramatic, possible effects of long-term exposure were often dismissed.

Along with measurements of atmospheric variables were empirical approaches for studying the human-atmosphere interface. Most important of these were those of the early seventeenth-century Italian experimenter Santoro Santorio, whose pioneering metabolic studies established the centrality of insensible perspiration—largely transdermal gaseous/vaporous output—as a determinant of health and disease. Santorio recognized that states of atmosphere affected such permeability, largely through facilitating or impeding evaporation from the skin. While Santorian explanations were popular, there were few efforts to extend them. In any case, here the atmosphere was important more for its capacity to remove—its emptiness—than for the effects of its components or positive conditions.

A final piece of background is the vehemence with which empiricism was being defended by the end of the eighteenth century. Sydenham's pioneering defiance and the seeming purity of neo-Hippocratism, contrasted with the ongoing promulgation of plausible but untestable pathophysiological systems, led French medical revolutionaries to seek to cleanse medical vocabulary of whatever was not measurable or observable.[21] Atmospheric variables were observable, yet they were not a key part of the Parisian agenda.

SURGEON REGINALD ORTON

On finally to Reginald Orton, whose study of cholera transcended the usual limits of atmospheric empiricism. Orton was a pioneering investigator of an apparently new disease for which there were no clear paradigmatic models that one was expected to emulate. Digesting an extraordinary range of medical and natural philosophical literatures, he arrived at generalizations about cholera that meet the four criteria identified above.

Orton was one of many middle-class British practitioners who saw the East India Company as a way to make a living and even as a means for self-betterment, in part through medical research. The surgeon Reginald Orton is distinguished only with difficulty from others with the same name. He was a younger son of Rev. Reginald Orton (d. 1802), rector of Hauxwell, Yorkshire, and his better-known nephew Reginald (also a surgeon) was a correspondent of Charles Darwin. Reginald's brother James was also a surgeon in the India Company. The relatively meager sum in their father's estate (£764) suggests the family was financially vulnerable. Reginald's letters from India to his mother and sisters are preoccupied with indebtedness and expense, with sending money home to support the family, and, ultimately, as regimental surgeon in the Thirty-Fourth (the Cumberlands), with raking in the sums that would allow retirement in England by age 30.[22] His medical training is not clear; that he was already in India as an assistant ship's surgeon at age 20 suggests that it came mainly through apprenticeship, perhaps supplemented by a year at Edinburgh.

The potential for medical fame from Indian investigations was especially high in the 1820s; the deadly world-threatening cholera, apparently new in 1817, came from India and could be understood only by those with Indian experience. Orton was only one of a number of medical men who took advantage of this unexpected authority. The second edition of his *Essay on Cholera* appeared in 1831, after his retirement to Hauxwell. The first edition (though only one of its two planned volumes) had appeared in 1820, published at Madras, intended primarily for circulation among the army's medical board. It appears not to have survived. The second edition (almost 500 pages), while never exactly popular, was highly respected.[23] Most of the changes in the second edition were included as a supplement, which, as reviewers pointed out, had a different theory of cholera. Over ten years, Orton had come to accept contagious transmission of cholera though he still upheld a geochemical origin.[24] James Johnson, the reigning British authority on tropical medicine, with considerable Indian experience, lauded Orton's book in his *Medico-Chirurgical Review* as "the very best which we have on the subject of cholera—at least as far as varied and laborious research is concerned."[25]

Orton's book was erudite, wide-ranging, and unusually ambitious. The work was comprehensive. Here I focus only on the four long chapters dealing with cholera's causes: these begin with an exploration of the internal pathological events that manifest as cholera, move on to meteorological causes, astronomical (sol-lunar) causes, and then proceed to electrochemical atmospheric causes, at which point they reconnect to proximate causes.[26] Reviewers overlooked and/or mildly rebuked these chapters on cause, which seemed (and seem) bizarre—they were more concerned with cholera's course and cure. One wrote: "We now approach some chapters of the work that may properly be passed over with a very brief notice, and we cannot but regret that the author did not himself greatly curtail the long and even not very interesting discussions therein contained." However, the reviewer continued: "We cannot follow Mr. Orton through the elaborate and ingenious analysis he introduces of the various symptoms of cholera, for the purpose of shewing the validity of the inferences just quoted; but we must, in justice to him, remark, that this part of his work is as interesting as it is conclusive, and we feel confident that those who were previously inclined to take a different view of the subject will confess themselves, after the perusal of the 'analysis' referred to, converted from their former opinions, and convinced of the accuracy of those here maintained."[27]

Orton arrived at an atmospheric explanation of cholera through a process of reasoning that incorporated all of the four criteria. He identified the particular constituents/characteristics responsible, stipulated mechanisms in a way that would guide experimentation, and offered epidemiological confirmation. All this is grounded in extraordinarily eclectic background reading on meteorological, natural philosophical, and biomedical subjects (remarkably, Orton did this research all the while bemoaning the limited library resources available to him in India).

His chain of reasoning begins with pathology/mechanism, or, as he called it in neo-Galenic terms, "the proximate cause." Appealing to Benjamin Brodie's experiments of 1811, in which circulation was temporarily kept going in decapitated rabbits, Orton argued that the other physiological effects—spasms, a rapid cooling, and suppression of urine—mimicked cholera. He inferred accordingly that cholera does not have a specific seat in a single organ, but represents rapid and systemic depression of nervous system function.[28] The subjective psychological symptoms of early cholera, ranging from "despondency" to "terror," confirmed a nervous-centric pathology. In itself this interpretation was not radical. Orton was making cholera consistent with mainstream fever pathology; he alluded to Boerhaave and to van Swieten's compendious commentaries on Boerhaave's aphorisms[29] and drew also on the well-known fever texts of John Huxham and William Cullen.

As he assessed the major individual symptoms in terms of this hypothesis, Orton appealed to many others as well. For example, again, following a common line of contemporary research (also one in which Brodie was active), he compared the pathology of cholera with the effects of poisons, many of which attacked the nervous system. He also explored analogies to other conditions in Cullen's class of "neuroses," which refers to primarily nervous system–related conditions.[30]

Orton, awed by India's hot and humid climate as were many Anglo-Indians, saw the atmosphere as the probable exciting cause of such a sudden and comprehensive effect. Framing the problem in Sydenhamian terms, he made clear that what had to be explained was the irregularity of such epidemic outbreaks despite the continuing presences of local causes that spread or exacerbated them. India provided a good laboratory for inquiry into such matters because of the regularity of its weather and the possibilities of systemic data-gathering on both medical and meteorological phenomena employing Orton's fellow regimental surgeons, distributed wherever the Company's army went.

A chapter on Indian meteorology establishes a link between cholera incidence and powerful thunderstorms. A chapter on sol-lunar influences further establishes the preponderance of such storms with the fullness of the moon. But by far the longest chapter treats the relation between atmospheric pressure, electrical state, storms, earthquakes, and epidemics, what Orton called the "primary remote cause" of epidemic cholera. Here Orton drew heavily on Noah Webster's *Brief History of Epidemics and Pestilential Diseases* of 1799.[31] There Webster argued that earthquakes and pestilences had a common geophysical cause, one possibly associated with comet flybys. Orton supplemented this earthquake-pestilence association with his cholera-storms-earthquakes association. For the human body, the most immediate pathological force generated by these atmospheric and geophysical events was high aerial electronegativity. From Luigi Galvani's research on the electrical character of nervous fluid (and with it the romantic conception of life as an electrical state), Orton concluded that atmospheric electronegativity was the source of lowered nervous energy of cholera victims.

But one might ask *how* this pathological cause operated, and Orton took this as an appropriate question to answer. It did so, Orton argued, primarily through electrochemical changes incident on respiration: that is, the exchange of gases in the lungs was also a reversal of charge—oxygen itself was simply a vehicle for transmission of an electrical state.[32] He then turned to the long series of bell-jar experiments begun more than a century earlier by Robert Boyle and continued by the newer generation of pneumatic chemists. These might be integrated with studies of atmospheric electricity to arrive at the physico-chemical determinants not only of survival but of the production of certain symptoms. By Orton's time, the experimenter could not only vary pressures, the focus of much of Boyle's work, but also alter the mixture of gases. These changes in turn would alter physical conditions like temperature, and in complicated ways, states of electronegativity. When Boyle exhausted bell jars containing animals, the subjects exhibited the symptoms of cholera. QED, declared Orton: "The analogy between the affection of the animals in this instance [experimental settings] and cholera cannot but strike the most inattentive reader. In fact, this simple experiment may probably be considered as a perfect instance of the simultaneous production of *cloud, rain,* and *cholera,* in miniature, from deficiency of electric fluid, bearing the strongest analogy to the process which takes place on a general attack of the epidemic."[33]

In sum, in Orton's view, the electronegativity of the air causes a deficit in nervous fluid, which manifests as cholera.[34] The process ought to operate on all organisms with similar physiology to humans, and it does, claimed Orton. Cholera, it is said, is epizootic as well as epidemic. Earlier, he had noted a common Indian disease in horses, known simply as "the gripes," which resembled cholera.[35] Orton followed with a long series of confirmatory anecdotes and expansions. The association of putrefying substances with cholera and other diseases, for example, reflected the electronegativity of putrefaction. The moon's conspicuous influence as a promoter of putrefaction also operated through its association with storms, and hence with electronegativity; moonlight itself is not harmful, Orton insisted—notwithstanding common opinion.

If all this were right, Orton believed that he had established "the train of causes and effects which prove the fact."[36] Orton was not sure it was right; and recognized the prevailing hostility to theory. He knew that his train of thought had taken him through unfamiliar territory; he had juxtaposed several traditions of induction that were respectable individually but had not been made to mesh. He apologized, admitting that he himself had perhaps already succumbed to a madness induced by the very tropical environment whose dangers he was recognizing: "A vertical sun poses a physical obstacle to intellectual exertions, which the critic, seated calmly at his desk under a happier sky, can form little idea of."[37]

Orton argued, however, that hypothesizing itself would advance the course of science even if particular hypotheses (perhaps his) were refuted.[38] The fiercest of the proto-positivist French clinicians would not have agreed for they despised all hypothesizing; yet much of the work he had assembled was unimpeachably empirical. But rather than being refuted, Orton's work would be marginalized, particularly the most ambitious and synthetic parts of it. He was denied the central gift of the scientific community: engagement.

Why? From a standpoint of comprehensiveness Orton had addressed the range of questions medical inquirers have wanted to ask, even if, at times, they have seen no plausible ways to answer. A simple answer is that Orton's pastiche of arguments didn't look like the sort of science a serious European researcher would do, just as Noah Webster's did not. European medical science had already become much more disciplinary; few aspired to integrate simultaneously the physiological laboratory, the empirical investigations of physical meteorology, and Sydenhamian epidemiology, using the *Encyclopedia Britannica*, old issues of the Royal Society's *Philosophical Transactions*, and a motley assortment of outdated and, perhaps, incompatible medical treatises as binding agents. For however impressive might be the range of his reading, Orton was not immersed in the literature of any single contemporary community of researchers. Leaving aside modern knowledge of the disease we call "cholera," we might be tempted to see Orton as projecting a sort of natural history and physical geography of respiratory physiology. In turn it is hard to imagine any of his peers reading as widely as he did or feeling as competent to take on such speculation.

But it is too easy to overexoticize the Ortonian synthesis both for his time and for ours. With regard to his time, his synthesis was audacious and bizarre, but not fundamentally outré. It was generally accepted that illnesses responded to seasons, places, and airs, exemplified by the intermittent fever that was coming to be known as "malaria," but also by the urban typhus that seemed empirically related to the deterioration of air produced in crowded places. Recognition of the atmosphere as a mixture of gases was still recent; new elements that might exist as gases, in isolated or combined form, were still being discovered; atmospheric electricity was relatively new; and the way any of these might interact with temperature, moisture, or other variables like gravity or magnetic field anomalies (being charted by the indefatigable Humboldt and his followers) was an open question. It was possible and plausible that subtle differences in atmosphere involving these variables did produce place-specific health effects. For our time, we may think of smog, of ozone, which would briefly fascinate some mid-nineteenth-century researchers, and all those health effects that go with being a "downwinder" in some respect. In short, it is not beyond the pale to project the high points of a counternarrative of medical history in which environmental determinants of health would not be isolated latter-day breakthroughs, but would belong instead to a trajectory, underrecognized by historians and environmental health researchers alike, which is in its own way at least as robust as those that have been recognized as the mainline of medical history.

Several factors, contingent and largely independent, may be posited for the dismissal of an atmospheric/climatic pathology and the failure to follow up Orton's initiative. Some have to do with Orton himself, others with the social and geographic setting of medico-climatic-atmospheric science, others with emerging views of the hierarchy of kinds and sites of knowledge, and still others with independent trajectories in medicine that diverted attention from the physical environment.

With regard to Orton himself: despite the investment of authority by James Johnson, a primary overseer of the assimilation of Anglo-Indian productions into the compendium of European medical knowledge, Orton failed as an advocate of his own ideas. He was ambivalent, openly critical of his own theory, unable to express clearly how the contagious character of cholera fit in with its pathology and climatic/atmospheric determinants.[39] He would come out of retirement at age forty-one to help meet military medical needs during the first British cholera outbreak of 1831–33, but he did not live the long life that might have allowed him to extend his studies.[40]

The matter of setting is more complicated, as historians of science who have studied metropole-provincial issues would expect. With regard to colonial medical science in particular, Mark Harrison has argued that during Orton's period a good many colonial and military practitioners anticipated or spearheaded key developments in medical theory and practice and acquired remarkable authority in their later careers, as they returned to Europe to retire or to take up teaching or administrative positions. They helped to globalize medicine by bringing at least some of the accumulated experience of the periphery to bear on the medicine of the metropole.[41] In some respects this would persist: there are ample instances into the twentieth century of middle-class practitioners returning to Britain after early retirement from careers as medical officers in India and leveraging their experience into high positions of professional authority.

And yet, not all forms of medical knowledge were translatable or useful. The understanding of health and disease in India never fully overlapped with that in England. What would become "tropical medicine" reflected a realization that particular diseases and broader factors affecting health really were foreign to European experience in many respects. Researchers who remained in India would sometimes express irritation with the obtuseness of newly arrived European experts who confidently expected to apply the knowledge they had gained from metropolitan lecture rooms, hospitals, and laboratories. Interestingly, the best-known feature of tropical medicine has been the prominence of parasitology, but had the term stabilized two generations earlier it might have been centered on climate. Enormously important to Europeans in the tropics, climate was one of those features that was, especially with the retreat of European malaria, increasingly irrelevant to medicine in the metropole. Indeed, for medical purposes, "climate" effectively became an attribute of otherness; it was a significant departure from the temperate places that were considered the norm. Accordingly, data-gathering enterprises relating climatic/atmospheric variables to morbidity or mortality, however fascinating from the vantage point of armchair imperialism, did not readily translate into specialist hospital appointments or even knowledge that could be usable to general practitioners or public health officers. The result, however, was that no matter how they fashioned themselves epistemically, those who studied cholera and airs could not win. To the degree that they were allowed to play at all, it was by themselves and in India, far away from the metropolitan centers of biomedical research.

Four Other Approaches

To illustrate the challenges of framing research on the health effects of atmosphere/climate, it is helpful to compare Orton's approach with the agendas represented by four other later researchers: Elisha Bartlett, Horatio C. Wood Jr., James Bryden, and H. W. Bellew. I treat these research agendas, which have to do with the status of various ways of knowing, in chronological order.

The first is a peculiar concomitant of the early nineteenth-century campaign to ground medicine in facts, the empiricist campaign. Though this medical ideology arose in Paris, it would spread to America during the 1830s through the many American students who journeyed to Paris to walk its hospitals and hear its great medical lecturers. An American, Elisha Bartlett (1804–55), would author its major (if belated) manifesto, his 1844 *Essay on the Philosophy of Medical Science*. Bartlett's commitment to understanding distinct diseases in terms of distinct lesions, and his desire to distinguish as clearly as possible the observation-based medicine pioneered in Paris from the speculative systems of the past, led him to erect a sharp divide between pathology and physiology. Efforts to understand pathological states as variants of normal physiological processes were not only egregiously speculative; they were, in Bartlett's view, fundamentally mistaken in principle. The pathological domain was not a variant of normal physiological processes but something wholly other. There was no physiological analogue to pus. Bartlett made his point with air. Plainly, bodies needed to respire; plainly too, their continued healthy function depended on other physical and chemical conditions, inputs and outputs. Chronic (and sometimes acute) deficits in any of these could kill. Yet these factors—which included the components/conditions of Orton's Indian atmospheres—had *nothing* to do with diseases, properly understood.[42] Though it would not go unchallenged, Bartlett's relegation would reappear in the health/disease distinction underwriting the bacteriology that would come to dominate by the end of the century: physiological aberrations might still register, but the bacteriological—and later, the viral—diseases would come to exemplify the concept of disease. They would be transitory invasions, not simply damaging states of relation between body and environment.

Second is the curious fate of "thermic fever." In 1871 Harvard's Boylston Prize was awarded to the young Philadelphia physician Horatio C. Wood Jr. (1841–1920) for a monograph on what he called "thermic fever"—what we would call heat stroke, what the French called *coup-de-soleil*.[43] Wood was making a point in dubbing the disease a "fever." Not only in terms of dangerously elevated temperatures but with regard to other classical fever symptoms, heat stroke or sunstroke counted as a fever, and was often diagnosed as some generic form of fever. It was one form of what classical medical writers called "ephemeral fever," which might be the mysterious "ardent fever" of the Hippocratic texts. Wood's synthesis, like Orton's, was multidisciplinary: it rested on his own clinical experience with laborers brought into the Pennsylvania hospital during a humid Philadelphia August, but also on Anglo-Indian clinical reports. Wood also utilized a substantial body of physiological research, largely by German researchers, but including some of his own work, on the body's means of heat regulation and on effects of heat on various functions. That tradition included the spine-severing experiments of Brodie that had been central for Orton's understanding of cholera. Wood, and others, were implying too, *contra* Bartlett, that the physiological study of the body's response to high environmental temperatures and its means of heat regulation was foundational to the understanding of the pathology of infectious or contagious fevers. Here environment was not incidental. Two meteorological variables, heat (sun) and humidity, combined with human activity, hydration, body covering, and other factors having to do with acclimatization and individual constitution, accounted for deadly acute febrile disease. If one followed Wood, matters of human interaction with atmosphere/climate would move from tropical curiosities to exemplars of pathology.

Wood's thermic fever fared better than Orton's cholera theory, but it did not become part of the mainstream. We do not call heat stroke "thermic fever"; the term had pretty well disappeared by 1920, and even before then had almost always been used in company with heat stroke or sunstroke. Usually (though not exclusively), we reserve "fever" for infectious diseases, even when these often operate by mechanisms that Wood and the German physiologists were exploring. That Wood was himself an experimenter may have augmented his stature; that he was clearly a member of the metropole, having an elite career in Philadelphia academic medicine, surely did too. However interesting Wood's research, both for the single condition of heat stroke and in explicating general problems of hyperpyrexia, the mainstream of medical research would follow an agenda that privileged the discovery of the agents of specific infectious diseases; environmental conditions were incidental. To be sure, each of the microbes had some form of extrabody habitat; but so did every other organism.

Last and together are two post-Ortonian Anglo-Indian researchers who took up the same question of trying to understand the patterns of Indian disease in terms of atmospheric/climatic variables. Both James L. Bryden (1833–80) and H. W. Bellew (1834–92) simplified Orton's project; they eschewed speculations about how atmospheric conditions incited pathological processes in order to focus wholly on establishing the fact that they did. Despite their overwhelming marshaling of facts, they too were marginalized. Hence it is hard to ascribe Orton's marginalization simply to his departure from strict canons of empiricism.

Bryden and Bellew represent what was then a new style of Anglo-Indian public medicine. After the establishment of comprehensive British rule following 1857, the imperial government would invest in a medical-statistical civil service. Building on the systematic observations of regimental, municipal, and prison medical officers, these statisticians could compile and compare data, even if they were rarely in a position to make data-informed policy. In many cases, their cholera studies were as rigidly restricted to facts as any Parisian positivist could possibly ask. Here I focus on Bryden's *Epidemic Cholera in the Bengal Presidency: A Report on the Cholera of 1866–68 and Its Relations to the Cholera of Previous Epidemics* (1869; he would supplement it in later years), and Bellew's *The History of Cholera in India from 1862 to 1881* (1885). Surgeon Bryden was "statistical officer" to the sanitary commissioner of the government of India; Deputy Surgeon-General Bellew had been the sanitary commissioner of Punjab. Both books are concerned almost wholly with the relation of four variables: time, place, rainfall, and cholera deaths (though the latter is sometimes broken down to distinguish white from black and officers from men). About the last 200 of Bryden's roughly 450 pages are appended tables—and tables make up a good deal of the preceding text. Bellew takes up nearly 900 pages, most of them summarizing the interrelationship of these variables year by year and region by region.

For both, positivism was a corrective to the enormous cholera literature written by European theorists who saw the disease only rarely but were sure they understood it. Only in India, where cholera was common, could its determinants—ultimately matters of place and climate for both Bryden and Bellew—be determined. Though Bryden worried that even statistics might be too weak a method to capture cholera's causes (for averages, which he called "secondary" facts, would obscure the pathological processes occurring in individual bodies), he nevertheless inferred general laws of cholera from the reams of data—chiefly, the existence of a wind, which transmitted a cholera poison (a germ) that was a product of filth interacting with certain environmental conditions.[44] Bellew, though intrigued by Bryden's hypothesis, worried that his colleague had strayed too far from the safe land of fact in indulging in gratuitous speculations about cholera agents. He "would wish to dismiss from the mind all previous reports and fancies, and to be guided solely by the evidence of the records." He had reorganized cholera statistics, he explained, after serving on a cholera commission, and recognizing that "the claims of humanity and the interests of science alike" were not being served simply by their compilation. "A real and fixed relation of cholera and weather has long been suspected to exist," he noted, "but of its nature nothing definite was known."[45] Only recently (and largely, since Bryden's efforts) had Indian mortality and meteorological data been adequate to explore the question.[46] In Bellew's view, these data confirmed that cholera, like malaria and influenza, was "a phenomenon of season"; the "sudden and simultaneous seizure" both in one place and in many was inconsistent with a narrowly transmission-based model.[47]

Hence, while Bellew and Bryden shared Orton's orientation—let us practice induction from the rich supply of fact that India affords—they were conspicuously, and increasingly, cautious: the baroque had given way to minimalism. Yet these martyrs of positivism were marginalized too, as completely as Orton had been. The coming of Robert Koch to India in 1883 would be reason for most of the medical world (and medical historians too) to stop worrying about Indian weather, though for another few decades Anglo-Indian practitioners would stubbornly mix the older studies with the new bacteriology. Koch arrived in India from Egypt. He found a characteristic microbe in the guts of those who had died of cholera, the feces of those who suffered from it, and, less certainly, in the waters they drank.

Koch's discoveries were not logically incompatible with atmospheric/climatic correlates or even determinates. For most contemporaries, however, such factors no longer mattered. Cholera was a matter of the presence of *Vibrio cholerae*, which could be cultured. It could, presumably, be avoided or destroyed. In essence, the philosophy was "don't worry about whether it will rain, just bring an umbrella." Sheldon Watts has emphasized the ideological basis of this view. If cholera were due to climate, the British Empire could not be held responsible.[48] Following relatively quickly, recognition of the role of mosquitoes in the transmission of malaria would strengthen that orientation. Thus within roughly three decades the exemplars of pathological tropical airs would be explained away by discoveries that would confirm a view in which environmental media had only incidental significance in etiology. At best, the environment was merely the place where the pathogen existed, the site of exposure, a poor proxy for what could be measured directly.

It has been tempting to see Koch's approach as exemplifying a comprehensive biomedical research agenda, embodying the criteria I noted earlier. That temptation may be especially strong when comparing his approach to merely correlative undertakings like Bryden's or Bellew's. The bacteriologist isolates a pathogen, which can then be subjected to laboratory investigation in a wide variety of ways, involving biochemistry, immunology, and genetics. Then, at the macroscopic level, its presence and absence can be charted against the occurrence of the disease in question. But to Bellew what was being lost by this narrow fixation on microbial agents was more prominent than what was gained. He wrote disparagingly of "the fashion . . . to neglect the physiological study of the human economy in its relation to cholera, and to divert the attention from the consideration of the true pathology and therapeutics of the disease, in the pursuit of various theories of specific germs, water contamination, contagious communication, &c., which not only fail to elucidate the etiology of cholera, but also fail altogether to explain the pathology of the disease, whilst, at the same time, nothing is done in the direction of establishing fixed principles for its therapeutic treatment."[49] The criticism may seem curious: we often think of the germ theory as exemplifying an etiological conception of disease. Bellew argued for greater attention to pathology and to factors that produced variable responses to pathogenic agents. He was right that such matters were often overlooked in the excited discovery of pathogens. He found worrisome that a lack of interest in pathological mechanisms implied a lack of concern with therapy and with the environmental and the social complexity that cholera in India manifests. He, like Orton, thought that host-specific factors, including atmospheric/climatic factors, might still be important determinants of the occurrence (and course) of the disease in India. Bellew did not see an agent-based conception of cholera as a route to its prevention but rather a high-handed attempt to impose a universal scientific solution on a complicated regional health problem. But he makes clear that however impatient he may be with Europeans who pronounce on Indian cholera, he is not rejecting agent-based theories so much as he is seeking to supplement them. They offer only partial truths.[50]

We can now see that it need not be "either/or." On the far side of revolutions in genetics, cell biology, and biochemistry, we are well-placed both to recognize the breadth of the gulf between external (even cosmic) environment and internal pathology, the ancient problem of macrocosm and microcosm, and to cross that gulf. The chemistry available to Sydenham or to Orton was vastly insufficient to the problem at hand. They could not fully know that, and each expressed unease differently. That so few felt comfortable pursuing the sorts of inquiries Reginald Orton did is illustrative, however, of interesting features of the division of intellectual labor, of the interplay of styles of scientific research, and, most profoundly, of the power of metropolitan laboratory science over provincial empiricism. These affected the conduct of climatic or geographic medicine in the nineteenth century and after. Those who, like Bryden and Bellew, limited themselves to the empirical found their achievements largely overlooked; that was true even of metropole-based synthesizers of medical climatology, like August Hirsch and Max von Pettenkofer.[51]

Contemporary cholera researchers now do exactly the kind of research Orton and his successors were proposing. Those interested in the ecology of *Vibrio cholerae* (and not merely in its status as cholera's agent) and who seek to understand its "waves" (a Brydenian term) have looked to the multiple factors that determine sea surface temperatures in the southern oceans, and as El Niño/La Niña-Southern Oscillation (ENSO), give rise to peculiar meteorological "constitutions" worldwide. One may now include under the heading of "mechanistic" not merely the way in which an agent operates within the body but the way in which the environment operates on that agent. Some have even begun to use the great nineteenth-century concordances of atmosphere-disease data to test their hypotheses, particularly Bellew's great induction of a three-year cholera cycle.[52] For Bellew himself, that cycle had simply been a "fact plainly declared." Repudiating the view that the cycle itself was an intellectual achievement, Bellew also resisted the temptation to seek its explanation: "I am not responsible. It is no invention of my own, nor do I pretend to offer an explanation of the phenomenon as such."[53]

So much for cholera. For air and climate more generally, however, it may be argued that analogous double standards remain in place—evident in a discomfort with recognizing what cannot be readily reduced to some single, simple, permanent, and well-defined entity. A modern Orton would probably struggle to be heard due to perceptions of his or her eclecticism; likewise a modern Bellew might be told that such integrative empiricism has little standing on its own and can be only suggestive. Beyond the confusion of correlation with cause, the question of what we mean by cause and whether causes like qualities of air can fully signify remains problematic, entangled in expectations about medical explanation that remain, to a large degree, artifacts of the historical processes discussed above.

NOTES

1. Harrison, "From Medical Astrology"; Mukharji, "Cholera Cloud."

2. Hamlin, *Cholera*, "Bacteriology," "Cholera Stigma," and "Environment and Disease."

3. For a review of literature see Valençius, "Histories of Medical Geography." Similar and overlapping issues arise with regard to occupational health. See Sellers, *Hazards of the Job*.

4. For a significant critique of this view see Harrison, *Medicine*.

5. Examples include Andrea Thompson, "Pollution May Cause 40 Percent of Global Deaths," Livescience, Sept. 10, 2007, accessed Nov. 1, 2013, http://www .livescience.com/1853-pollution-40-percent-global-deaths.html; Tom Levitt, "Air pollution linked to 200,000 premature deaths in U.K," *Ecologist*, Dec. 21, 2010, accessed Nov. 1, 2013, http://www.theecologist.org/News/news_round_up/701026/ air_pollution_linked_to_200000_premature_deaths_in_uk.html; Joanna Zelman, "Power Plant Air Pollution Kills 13,000 People Per Year," *Huffington Post*, May 25, 2011, accessed Nov. 1, 2013, http://www.huffingtonpost.com/2011/03/14/power-plant -air-pollution-coal-kills_n_833385.html; and "In News: Mortality and Morbidity due to Air Pollution," accessed Nov. 1, 2013, http://urbanemissions.info/sim-series-06/ mortalityandmorbidity.html.

6. American Lung Association, "What's the State of Your Air?" last modified Jan. 4, 2012, http://www.stateoftheair.org/.

7. American Lung Association, "2011 State of the Air: Key Findings," http://www .stateoftheair.org/2011/key-findings/.

8. World Health Organization, "Air Quality and Health: Fact Sheet No. 313," last modified Sept. 2011, accessed Jan. 6, 2012, http://www.who.int/mediacentre/ factsheets/fs313/en/index.html.

9. Regarding waterborne cancers, see Nash, *Inescapable Ecologies*.

10. Recent examples include Porter, *Greatest Benefit*, and W. F. Bynum, *Science and the Practice of Medicine*. Harrison, *Disease*, discusses dangerous places, but does not analyze specific climatic/atmospheric factors.

11. Latour, *Pasteurization of France*.

12. Said, *Orientalism*.

13. Riḍwān, *Medieval Islamic Medicine*; Sharma and Dash, *Agniveśi's Caraka-Samhitā*; Veith, *Huang Di Nei Jing Su Wên*.

14. Macculloch, *Malaria* and *Essay on the Remittent and Intermittent Diseases*.

15. Porter, *Greatest Benefit*, 302.

16. Hill, "Environment and Disease," examines standards of causal inference.

17. Sydenham, *Medical Observations*, sections 2.5, 2.22, pp. 33, 40.

18. Riley, *Eighteenth-Century Campaign*.

19. Golinski, *British Weather*.

20. Schaffer, "Measuring Virtue."

21. Staum, *Cabanis*, and Foucault, *Birth of the Clinic*.

22. "Reginald Orton, Military surgeon, India," accessed Jan. 6, 2012, http://free pages.history.rootsweb.ancestry.com/~enzedders/roletters.htm; "Reginald Orton, Rector of Hauxwell [Yorkshire]," accessed Jan. 6, 2012, http://freepages.history .rootsweb.ancestry.com/~enzedders/ortonwills.htm#21; and Power, "Reginald Orton (1810–1862)."

23. Orton, *Essay on the Epidemic Cholera*.

24. Ibid., xiv-xvi.

25. Johnston, *Medico-Chirurgical Review*, 211.

26. The parts I am concerned with are from the first part—the 1820 first edition.

27. Unsigned review of *An Essay on the Epidemic Cholera of India*, by Reginald Orton, *London Medical and Physical Journal* n.s. 12 (1832): 44.

28. Orton, *Essay*, 58–60; cf. [Brodie], "Medical Extracts."

29. Swieten, *Commentaries*.

30. Orton, *Essay*, 133–57.

31. Webster, *Brief History*.

32. Orton is drawing on Read, "Experiments and Observations," 423–26.

33. Orton, *Essay*, 285–86.

34. Ibid., 298, 301.

35. Ibid., 156.

36. Ibid., 301.

37. Ibid., xi.

38. Ibid., viii–x; cf. 56–57.

39. See Orton, *Essay*, 313–47.

40. Orton, "Observations," 41–43.

41. Harrison, *Medicine in an Age of Commerce*.

42. Bartlett, *Essay*; Ackerknecht, "Elisha Bartlett"; and Warner, *Against the Spirit*.

43. Wood, *Thermic Fever*.

44. Bryden, *Epidemic Cholera*; Harrison, "Question of Locality"; *Public Health*, 100–101.

45. Bellew, *History of Cholera*, 15–16.

46. Ibid., ix-x, 2–4, 11–14.

47. Ibid., 9, 13–14.

48. Watts, "From Rapid Change to Stasis" and "Cholera Politics." But see Harrison, *Public Health*, 109–16.

49. Bellew, *History of Cholera*, ix.

50. Ibid., ix, 5.

51. Cf. Hirsch, *Handbook*, with Davidson, *Geographical Pathology*.

52. Lipp et al., "Effects of Global Climate"; Pascual et al., "Cholera and Climate."

53. Bellew, *History of Cholera*, v.

REFERENCES

Ackerknecht, Erwin Heinz. "Elisha Bartlett and the Philosophy of the Paris Clinical School." *Bulletin of the History of Medicine* 24 (1950): 43–60.

Bartlett, Elisha. *An Essay on the Philosophy of Medical Science*. Philadelphia: Lea and Blanchard, 1844.

Bellew, H. W. *The History of Cholera in India from 1862 to 1881*. London: Trübner, 1885.

[Brodie, Benjamin]. "Medical Extracts." *Edinburgh Medical and Surgical Journal* 8 (1812): 447–52.

Bryden, James L. *Epidemic Cholera in the Bengal Presidency: A Report on the Cholera of 1866–68 and Its Relations to the Cholera of Previous Epidemics*. Calcutta: Office of Superintendent of Government Printing, 1869.

Bynum, W. F. *Science and the Practice of Medicine in the Nineteenth Century*. Cambridge: Cambridge University Press, 1994.

Bynum, W. F., Anne Hardy, Stephen Jacyna, Christopher Lawrence, and E. M. Tansey. *The Western Medical Tradition: 1800 to 2000*. Cambridge: Cambridge University Press, 2006.

Davidson, Arnold. *Geographical Pathology*. Edinburgh: Young J. Pentland, 1892.

Foucault, Michel. *The Birth of the Clinic: An Archaeology of Medical Perception*. Translated by A. M. Sheridan Smith. New York: Pantheon, 1973.

Golinski, Jan. *British Weather and the Climate of Enlightenment*. Chicago: University of Chicago Press, 2007.

Hamlin, Christopher. "Bacteriology as a Cultural System: Analysis and Its Discontents." *History of Science* 49, no. 3 (2011): 269–98.

———. "Environment and Disease in Ireland." In *Environment, Health, and History*, edited by Virginia Berridge and Martin Gorsky, 45–68. London: Palgrave Macmillan, 2011.

———. "The Cholera Stigma and the Challenge of Interdisciplinary Epistemology: From Bengal to Haiti." *Science as Culture* 21, no. 4 (2012): 445–74.

———. *Cholera: The Biography.* Oxford: Oxford University Press, 2009.

Harrison, Mark. "From Medical Astrology to Medical Astronomy: Sol-lunar and Planetary Theories of Disease in British Medicine, c. 1700–1850." *British Journal for the History of Science* 33 (2000): 25–48.

———. "A Question of Locality: The Identity of Cholera in British India, 1860–1890." In *Warm Climates and Western Medicine: The Emergence of Tropical Medicine,* edited by David Arnold, 133–59. Amsterdam: Rodopi, 1996.

———. *Disease and the Modern World, 1500 to the Present Day.* London: Polity, 2004.

———. *Medicine in an Age of Commerce and Empire: Britain and Its Tropical Colonies, 1660–1830.* Oxford: Oxford University Press, 2010.

———. *Public Health in British India: Anglo-Indian Preventive Medicine, 1859–1914.* Cambridge: Cambridge University Press, 1994.

Hill, Austin Bradford. "The Environment and Disease: Association or Causation?" *Proceedings of the Royal Society of Medicine* 58, no. 5 (1965): 295–300.

Hirsch, August. *Handbook of Historical and Geographical Pathology.* 2nd ed. London: New Sydenham Society, 1883.

Johnston, James. *Medico-Chirurgical Review* 31 (Jan. 1832): 194–211.

Latour, Bruno. *The Pasteurization of France.* Translated by Alan Sheridan and John Law. Cambridge, MA: Harvard University Press, 1988.

Lipp, E. K., A. Huq, and R. R. Colwell. "Effects of Global Climate on Infectious Disease: The Cholera Model." *Clinical Microbiology Reviews* 15, no. 4 (2002): 757–70.

London Medical and Physical Journal. Unsigned review of *An Essay on the Epidemic Cholera of India,* by Reginald Orton. N.s. 12 (1832): 42–52.

Macculloch, John. *An Essay on the Remittent and Intermittent Diseases Including, Generically, Marsh Fever and Neuralgia . . . .* 2 vols. London: Longman, Rees, Orme, Brown, and Green, 1828.

———. *Malaria, An Essay.* London: Longman, 1827.

Medicina Statica Hibernica, or, Statical Experiments to examine and discover the insensible perspiration of a Human Body in the south of Ireland. Dublin, 1734.

Mukharji, Projit B. "The 'Cholera Cloud' in the Nineteenth-Century 'British World': History of an Object-without-an-Essence." *Bulletin of the History of Medicine* 86, no. 3 (2012): 303–32

Nash, Linda. *Inescapable Ecologies: A History of Environment, Disease, and Knowledge.* Berkeley: University of California Press, 2006.

Orton, Reginald. "Observations on the Malignant Cholera in England." *Lancet* ii (1833): 41–43.

———. *An Essay on the Epidemic Cholera of India.* 2nd ed. London: Burgess and Hill, 1831.

Pascual, M., M. J. Bouma, and A. P. Dobson. "Cholera and Climate: Revisiting the Quantitative Evidence." *Microbes and Infection* 4, no. 3 (2002): 237–45.

Porter, Roy. *The Greatest Benefit to Mankind: A Medical History of Humanity.* New York: Norton, 1998.

Power, D'A., and Mark Clement. "Reginald Orton (1810–1862)." In *Oxford Dictionary of National Biography.* Oxford University Press, 2004–. Accessed Jan. 6, 2012. doi:10.1093/ref:odnb/20857.

Read, John. "Experiments and Observations made with the Doubler of Electricity, with, a View to Determine its Real Utility, in the Investigation of the Electricity of Atmospheric Air, in Different Degrees of Purity." In *Philosophical Transactions of the Royal Society of London, from their Commencement, in 1665, to 1800. Abridged*, edited by Charles Hutton, George Shaw, and Richard Pearson. Vol. 17, 1791 to 1796, 423–26. London: C. and R. Baldwin: 1809.

 Riḍwān, Ali ibn, Michael W. Dols, and Jamal Ādil Sulaymān. *Medieval Islamic Medicine: Ibn Riḍwān's treatise, "On the prevention of bodily ills in Egypt."* Berkeley: University of California Press, 1984.

Riley, James C. *The Eighteenth-Century Campaign to Avoid Disease.* London: Macmillan, 1987.

Said, Edward. *Orientalism.* New York: Random House Vintage, 1979.

Schaffer, Simon. "Measuring Virtue: Eudiometry, Enlightenment, and Pneumatic Medicine." In *The Medical Enlightenment*, edited by Andrew Cunningham and Roger French, 281–331. Cambridge: Cambridge University Press, 1990.

Sellers, Christopher. *Hazards of the Job: From Industrial Disease to Environmental Health Science.* Chapel Hill: University of North Carolina Press, 1997.

Sharma, Ram Karan, and Vaidya Bhagwan Dash. *Agniveśi's Caraka-Samhitā.* 2 vols. Chowkhamba Sanskrit Studies 94. Varanasi: Chowkhamba Sanskrit Series Office, 1977.

Staum, Martin. *Cabanis: Enlightenment and Medical Philosophy in the French Revolution.* Princeton, NJ: Princeton University Press, 1980.

Swieten, Gerhard van. *Commentaries upon Boerhaave's aphorisms concerning the knowledge and cure of diseases.* 15 vols. Edinburgh: Charles Eliot, 1776.

Sydenham, Thomas. *Medical Observations concerning the History and Cure of Acute Diseases.* London: Sydenham Society, 1844.

Valençius, Conevery Bolton. "Histories of Medical Geography." In *Medical Geography in Historical Perspective*, edited by Nicholas Rupke, 1–28. London: Wellcome Trust Centre for the History of Medicine, 2000.

Veith, Ilza. *Huang Di nei jing su wên. The Yellow Emperor's Classic of Internal Medicine*. Berkeley: University of California Press, 1966.

Warner, John Harley. *Against the Spirit of System: The French Impulse in Nineteenth-Century American Medicine*. Princeton, NJ: Princeton University Press, 1998.

Watts, Sheldon. "Cholera Politics in Britain in 1879: John Netten Radcliffe's Confidential Memo on 'Quarantine in the Red Sea.'" *Journal of the Historical Society* 7, no. 3 (2007): 291–348.

———. "From Rapid Change to Stasis: Official Responses to Cholera in British-Ruled India and Egypt, 1860 to c. 1921." *Journal of World History* 12, no. 2 (2001): 321–74.

Webster, Noah. *A brief history of epidemic and pestilential diseases with the principal phenomena of the physical world, which precede and accompany them, and observations deduced from the facts stated*. 2 vols. Hartford: Hudson & Goodwin, 1799.

Wood, Horatio C., Jr., *Thermic Fever, or Sunstroke*. Philadelphia: Lippincott, 1872.

Zysk, Kenneth G. "Fever in Vedic India." *Journal of the American Oriental Society* 103, no. 3 (1983): 617–21.

3

BETTER TO CRY THAN DIE?
THE PARADOXES OF TEAR GAS IN THE VIETNAM ERA

Roger Eardley-Pryor

TEAR GAS, A nonlethal chemical weapon typically used for riot control, presents several paradoxes. Tear gas is not a gas but a micropulverized powder that causes uncontrollable tears, irritated breathing, and escalating pain when inhaled in sufficient concentrations. Although classified as nonlethal, powerful tear gases burn skin, induce nausea, and, though uncommon, can result in death by asphyxiation.[1] Tear gas's greatest paradox, however, may be that American police forces face few regulations on its use, and citizens can purchase military-grade supplies over the internet, yet international protocols forbid the use of tear gas as a weapon in war, despite its proven ability to limit deaths. Why is tear gas considered globally a humane *domestic* police tool but forbidden as a weapon in *international* warfare?

Paradoxical domestic and international policies for tear gas emerged in the United States during the Vietnam era, between the mid-1960s and the early 1970s. US soldiers applied nearly fifteen million pounds of tear gas during the Vietnam War—enough chemical to coat all of Vietnam with a field-effective concentration. In the first years of US military deployments in Vietnam, two justifications prevailed for tear gas use in war: what I call the home argument and the humane argument. The home argument referenced domestic use of nonlethal chemicals by police forces, especially as a technological fix for domestic race riots then sweeping the nation.[2] The humane argument identified tear gas as a compassionate technology that saved lives rather than ended them. But after American troops launched tear gas to flush Vietcong out of tunnels and kill them, the military abandoned the humane argument in the face of public opposition and advocated a tactical argument for tear gas in war.[3]

By 1970, political opinion had shifted against using nonlethal chemical weapons in war. The burgeoning environmental movement's general consternation over chemicals and a series of alarming incidents involving lethal chemical weapons dovetailed with growing disapproval of American actions in Vietnam, including the use of chemicals to flush out and kill people.[4] These chemical anxieties coalesced into support for what might be called a slippery slope argument, which eventually overpowered the home, humane, and tactical arguments for tear gas in war. The slippery slope argument identified nonlethal chemicals as dangerous gateway agents to be banished from international combat before the proliferation and reintroduction to war of lethal chemical weapons.

Most studies of chemical weapons focus on lethal chemical agents, or, as in the case of Vietnam, on chemical herbicides.[5] This chapter examines why Americans in the Vietnam era initially endorsed nonlethal tear gas in war, but eventually banned it from international combat, all while considering tear gas a necessary and legal tool at home. It explains how paradoxical policies for this chemical weapon developed after domestic and international uses intersected with complex racial, environmental, and antiwar social movements. The shifting status of tear gas also coincided with growing skepticism about the promises of science and technology during this era.[6] Due to such skepticism and to the American military's excessive use of tear gas in Vietnam, American policymakers eventually prohibited the use in war of a nonlethal chemical that domestic police have used freely since the early twentieth century. In April 1975, President Gerald Ford officially renounced as a matter of US policy the first-use of tear gas in war except in very prescribed circumstances.[7] This new policy only restricted the military's tear gas applications in combat; it had no bearing on the use of tear gas by domestic police forces against American citizens.

Chemical Nonviolence?

German scientists first synthesized the tear-inducing chemical CN (choloracetophenone) in the late nineteenth century. During World War I, all belligerent nations, including the United States, weaponized CN and launched tear gas munitions in battle before they escalated to lethal chemical weapons. Following WWI, veterans of the US Army's Chemical Warfare Service (CWS) encouraged domestic police forces to adopt military tear gas as a nonviolent tool to control widespread labor unrest and race riots.[8] Between 1919 and 1921 alone, at least twenty-nine violent strikes and riots necessitated federal military intervention to restore civil order. By the end of 1923, police in six hundred cities possessed tear gas and celebrated its effective use against civilians. Only a few years after WWI, the United States had adopted tear gas as a civilian technology.[9]

As domestic use of nonlethal chemical weapons spread in the United States, international opposition to deadly chemical weapons led several nations to attempt to prohibit their use in future wars. The Geneva Protocol of 1925 proposed banning "the use in war of asphyxiating, poisonous or other gases and all analogous liquids, materials or devices." A successful lobbying effort by the CWS and the US chemical industry painted gas warfare as potentially humane, which ultimately kept the US Senate from ratifying the protocol and made the United States the world's only industrialized military power not to have ratified it. Over the following decades, unofficial US policy rejected the first-use of chemical weapons in war, even while supplying US troops with tear gas.[10]

The Cold War arms race, however, encouraged drastic expansion of classified research, development, and production of highly lethal chemical and biological weapons. Tear gas benefited from this expansion when, in 1958, the US Army replaced CN as the standard riot control agent with the more powerful tear gas CS (o-chlorobenzylidene malononitrile), named for the two American chemists, Corson and Stoughton, who first synthesized it. Most domestic police forces followed suit.[11] Years of use of tear gas in domestic uprisings inspired American military planners in the 1960s to adopt tear gas for counterinsurgency situations abroad. The United States' failure to ratify any treaties banning chemical weapons in war encouraged military experimentation with tear gas in southeast Asia.

In 1962, the United States began equipping the South Vietnamese Army with nonlethal riot control gases. In November 1964, they used tear gas to quell a Buddhist revolt and, with US encouragement, secretly initiated its use against Vietcong insurgents. In March 1965, US Marines engaged for the first time in combat in Vietnam where they secretly used tear gases and chemical herbicides. Before the month's end, the quiet use of such nonlethal chemical agents erupted in a political firestorm, catching President Lyndon Johnson's administration off-guard.[12]

OVERSTEPPING THE SLIPPERY SLOPE

Almost immediately the American, British, German, Japanese, and communist press in Peking and Hanoi deplored the American reintroduction of gas warfare, while overlooking the chemicals' nonlethality. The Japanese paper *Asahi* accused the United States of using Asians as guinea pigs for chemical experiments, while the communist press called Americans fascist cannibals. A Frankfurt newspaper published a cartoon depicting the Statue of Liberty wearing a gas mask, and another Japanese paper carried a cartoon of Adolf Hitler's ghost hovering over Vietnam holding a gas-inflated bag.[13] President Johnson demanded that McGeorge Bundy, his National Security Advisor, explain what Bundy discarded as a "stupid fuss over gas." In a memo, Bundy outlined the home and humane arguments, which Johnson then marketed to the United States and international media.[14]

In his news conference of April 1, 1965, President Johnson portrayed tear gases in Vietnam as humanitarian, harmless, and the same as those used by domestic police forces. He declared, "where women and children are involved, where American citizens are involved, where the enemy is involved ... it is not always the better course of wisdom to handle that situation with machine guns or bombs or implements of war that bring death." He described tear gas as a standard "riot-control item" that any American could purchase, "just like you order something out of a Sears Roebuck catalog." Johnson affirmed that police across the nation used the same nonlethal gases. In fact, he continued, "the Chief of Police in Washington [DC] has it now and if in the interest of saving lives and protecting people, it would be used."[15]

Despite the nonlethal nature of tear gases, critics castigated American reintroduction of a military activity that had been almost universally condemned since the end of WWI. Many scientists and arms control advocates claimed consistently that use in war of nonlethal chemical agents would set the world down the slippery slope of a chemical arms race, ending potentially in massive deaths and environmental devastation from relatively cheap and highly lethal chemical weapons. A working group at the Fourteenth Pugwash Conference in April 1965 advocated banning all chemical weapons spanning "a continuous spectrum, starting from those that are temporarily incapacitating and ending with highly lethal ones." They warned that "the entire spectrum of these weapons may come into use" if restraints on their use broke down, like applying tear gas in combat.[16] Critics suggested that, once gases were introduced, unstable developing nations might produce and use cheap but deadly chemicals, removing American advantages of superior firepower. For the sake of global security, slippery slope advocates believed that all chemical weapons, even if nonlethal, must be prohibited from warfare.

Though clear sides were drawn, Johnson's two-pronged justification, that tear gas was humane and harmless, initially triumphed. America's British allies voiced relief at Johnson's explanations. Upon learning that British armed forces abroad had used tear gas over 120 times in the previous five years, a Conservative member of Parliament replied, "While most of us, I think, disagree with the slogan, 'Better Red than dead,' I think we would all agree that it's better to cry than die."[17] Some scientists concurred. A microbiologist, Julian Pleasants, writing in *Commonweal*, a prominent Catholic journal, believed chemical weapons helped warfare "evolve in the direction of humaneness." Pleasants connected tear gas use abroad to domestic use by asking, "Would it have been less evil for Southern police to machine-gun Freedom Marchers than to tear-gas them?" Using minimum force via nonlethal chemicals, he argued, made war more moral.[18] Despite waning public opposition, the Johnson administration suspended further tear gas use by US forces as a precaution. Press reports of gas in Vietnam virtually disappeared.[19] Like a cloud of tear gas on a windy day, the uproar about tear gas in Vietnam dissipated—at least temporarily.

ESCALATION ABROAD

A few months later, in September 1965, Saigon announced that US commander Col. L. N. Utter faced disciplinary charges for the unauthorized use of tear gas in Vietnam. Col. Utter had used nearly fifty canisters of tear gas to flush out and separate several hundred villagers from suspected Vietcong insurgents. According to the *New York Times*, Utter "felt that tear gas was the most humane way to dislodge the Vietcong suspects." Although Utter faced serious charges, the press celebrated this humane application of gas: "All emerged from their hideaways weeping and sniffing, but they were otherwise unharmed."[20] The Utter trial provided a public-reaction test for America's reintroduction of tear gas to war.

After the press reported the Utter episode without public opposition—even amid accusations that several civilians died during the incident—US tear gas use in Vietnam escalated rapidly. One month after Utter's full exoneration, US field commanders received authorization to use CN and CS

FIG. 3.1. Helicopter dropping an exploding fifty-five-gallon drum of tear gas over a suspected Vietcong stronghold. SOURCE: Photograph No. 111-CCV-104-CC41406; Photographs of US Army Operations in Vietnam, compiled 1963–1973, Record Group 111, National Archives at College Park, College Park, MD.

gases as they would most any other weapon.[21] New military tools expanded tear gas's routine application, including massive air drops from helicopters (see fig. 3.1) and the use of "Mitey Mite," a portable, gasoline-powered fan that spread pounds of crystallized CS at high speeds to keep tunnels and other areas unlivable for long periods. Tear gas victims, the American press crowed, supposedly suffered "no lasting effects."[22] On the other hand, some antiwar scientists warned that "even if the war in Vietnam is limited to tear gas, other countries in other wars might be encouraged to use gas in a more general way."[23]

Early in 1966, the *New York Times* front page reported the routine application of tear gas to flush out embedded Vietcong troops immediately before B-52 air raids carpet-bombed the weeping enemy. Reaction to the story was muted. Body counts mattered most now, not the mere capture of combatants with the use of nonlethal chemicals. In March 1966, the *New Repub-*

lic claimed definitively that "we have deliberately crossed the long-observed [chemical and biological] warfare threshold . . . And we have done it with barely a ripple of protest . . . from any but Communist nations."[24] The lack of significant protest signaled broad acceptance of the humane argument, ironically, even when ground troops used tear gas tactically to flush enemy troops from their hiding places in order to kill them with other methods (see fig. 3.2).

In September 1966, twenty-two American scientists, including seven Nobel laureates, asked President Johnson to halt use of all chemical weapons in Vietnam. The scientists repeated their slippery slope argument: "If the restraints on the use of one kind of [chemical or biological] weapon are broken down, the use of others will be encouraged."[25] The scientists garnered some headlines, but the Johnson administration's tacit reply appeared in a dramatic escalation of chemicals for combat. According to records from the Institute for Defense Analyses, the levels of CS gas procurement by the Department of Defense rose by a factor of six from 1965 to 1966, and it quadrupled from 1966 to 1968.[26] Over the same period, the rationale for tear gas use in war gained further reinforcement from domestic events.

Domestic Disorder

In the summer of 1965, race riots in the Watts neighborhood of Los Angeles erupted unexpectedly and, over the next few years, reverberated across urban America with widespread street violence, firebombing, looting, and sniper fire. A surge in race-related riots and vast social unrest followed, shaped in part by the escalating war in Vietnam. In the hot summer of 1967, both Newark and Detroit suffered catastrophic riots before the National Guard and state police regained control. By the end of the summer, over one hundred cities experienced incidents of racial violence—from Rochester, New York, to Portland, Oregon; from Milwaukee, Wisconsin, to Riviera Beach, Florida—with over one hundred people killed, thousands injured, and property damage estimated around $1 billion measured in 1967 dollars. The entire nation was in turmoil.[27]

These events led concerned citizens, black and white, to recognize tear gas as a reasonable substitute to lethal weapons for reestablishing social order. The National Chamber of Commerce explained that "to preserve our free way of life in a reasonably orderly society, it is essential that riots be controlled without violence." New York Police Commissioner Howard Leary explained, "We must do more to explore and develop non-deadly weapons. The gun is not the way to prevent most crime." A public safety advisor to the

FIG. 3.2. In 1968 a US Marine, gas mask on, waits outside a Vietcong tunnel sprayed with tear gas. SOURCE: Photograph No. 127-GVB-83-A192216. Black and White Photographs of Marine Corps Activities in Vietnam, compiled 1962–1975, Record Group 127, National Archives at College Park, College Park, MD.

Department of Housing and Urban Development in Washington declared, "The civil disorders this summer that overwhelmed the capabilities of many police departments demonstrate the need for modern nonlethal weaponry to deal with crowds and mobs." Journalists covered nonlethal technologies extensively as solutions to domestic violence, with many referencing the use of tear gas in Vietnam.[28]

Rather than exploring social or political remedies for curbing civil unrest, America's elites sought technological solutions. Detroit's police commission even predicted, "we will be able to solve or prevent most crimes with computers ten years from now."[29] In the meantime, tear gas seemed to offer an immediate technological fix to violence both at home and abroad. The April 12, 1968, cover of *Time* magazine captured concerns about the war abroad and civil unrest at home with a sorrowful image of President Johnson under the words: "The Search for Peace in Asia, the Specter of Violence at Home."[30] In conjunction with its success in Asian battlefields, tear gas became the most common technological solution for riot control at home.

The FBI concurred. After the Detroit uprising and as part of the President's Commission on Law Enforcement and Administration of Justice, the FBI reissued its guidelines of 1965 on preventing and dealing with riots. To maintain civil order, the document reiterated the state's power to "use firearms and other auxiliary weapons of any kind," especially "tear gas . . . available commercially." When properly employed, it explained, nonlethal chemicals provided "the most effective and humane means of achieving temporary neutralization of a mob with a minimum of personal injury." Tear gas was so effective that the FBI cited gas masks as its primary recommendation of protective equipment for police mob squads, above helmets, body armor, or even boots. Domestic use of gas was not just humane and effective; it was also economical. "Skillful and prudent use of force," the FBI concluded, "will . . . accomplish the mission with minimum expenditure of resources."[31] Domestic events clearly crystallized the home argument for tear gas in combat, even if deadly tactical use of tear gas in Vietnam superseded the humane argument.

During the Tet Offensive in Vietnam in 1968, tear gas was used as both a defensive and offensive tool. In response to the Vietcong assault, particularly in the city of Hué, American forces employed large quantities of CS gas, celebrating its ability to provide protection in urban warfare and help flush out and then kill the enemy. The November 1968 edition of *Army Digest* explained that "entrenched areas that have successfully resisted both aerial and artillery fire have been reduced in an hour or two by combining the use of CS with maneuver and firepower."[32] The humane argument for tear gas in combat had evaporated.

American tear gas use skyrocketed at home and abroad. In 1968 US troops in Vietnam used almost twice as much CS gas than in all prior years combined; and in 1969 military use increased another 20 percent. According to figures collected by New York Congressman Richard McCarthy, the 13.7 million pounds of CS gas procured for use in Vietnam was enough, when dispersed by helicopter, to cover some 80,000 square miles. South Vietnam was only 66,000 square miles. On the home front, congressional passage of the Omnibus Crime Control and Safe Streets Act of 1968 provided grant money to state and local police forces for procurement and training in tear gas application.[33] In a climate of international and domestic insecurity, the federal government employed tear gas as a technological solution to both social and military problems.

WINDS OF CHANGE

Despite enthusiasm for tear gas in America's military and civilian security circles, by 1970 a growing environmental consciousness, combined with a series of revelations involving deadly chemical agents and the horrific deaths of students during peaceful campus protests, began to challenge America's unabashed use of nonlethal chemical weapons both at home and in war. A few years after Rachel Carson's publication of *Silent Spring* in 1962 helped inspire a new and vocal environmental movement, apprehension about chemical threats to the environment dovetailed with growing antiwar sentiments.[34] Renewed attention to chemicals in combat offered opportunities for concerned scientists and policymakers to shape national policy on chemical weapons.[35] A succession of incidents supported this trend against the use of chemical agents. The first came from the Red Cross's medical mission to Yemen in 1967, which confirmed suspicions about Egypt's use of poison gases in the war between Saudi Arabian-supported Yemeni royalists and the Egyptian-supported rebels. After the war had stalemated, Egypt experimented with tear gas attacks before escalating to the use of lethal gases, killing hundreds.[36]

The following year a deadly domestic chemical weapons accident seized the attention of policymakers, scientists, and the American media. Dugway Proving Ground, then the US military's primary site for testing chemical and biological weapons, lies twenty-five miles outside Skull Valley, Utah. In March 1968, shortly after a Dugway Air Force jet sprayed over three hundred gallons of highly lethal VX nerve gas in an open-air test, some six thousand sheep grazing in Skull Valley suddenly began writhing, foaming at the mouth, and dying grotesquely. Unexpected wind gusts had carried the VX gas nearly fifty miles from its target, with Salt Lake City only eighty miles away. The Army initially denied responsibility for the sheep's deaths, but, following a US Public Health Service investigation, the Army paid over half a million dollars in damages to the ranchers and lent bulldozers to bury the sheep carcasses. A local doctor, fearing future accidents, proclaimed, "The sheep's troubles are over. Now I'd start worrying about people."[37] No humans suffered injury, but the harrowing event captured significant media attention and brought home the serious threats of lethal chemical weapons for many Americans.[38]

Also in 1968, the investigative reporter Seymour Hersh published an array of articles in support of his book *Chemical and Biological Warfare*, a shocking treatise on America's mostly secret but extensive lethal chemical weapons capabilities. Hersh's book unveiled military secrecy surrounding the research and production of America's stockpiles of killer chemicals. Images of Skull Valley's dead sheep, piled by the thousand, accompanied Hersh's articles and reviews of his book.[39] As the Vietnam conflict settled into stalemate, Hersh had raised awareness and understanding of deadly chemicals.

American scientists concerned about the dangers of chemicals, generally, and chemical weapons, especially, conflated chemical threats to human health, the environment, and national security. After the Dugway sheep incident, Harvard biochemist John T. Edsall initiated roundtable discussions on the use of herbicides and tear gas in Vietnam. Edsall wondered whether use of nonlethal chemical weapons opened the door "to further escalation of their use by our enemies now or by other people later on that may lead to the use of enormously deadly nerve gases, which are really weapons of mass-destruction." J. B. Neilands, a biochemist at the University of California, chastised the wartime use of defoliants and tear gas as a breach of international law. He demanded US ratification of the Geneva Protocol of 1925, arguing that ongoing use of any chemicals in combat "could open a Pandora's Box leading ultimately to the acceptance and use of ever more toxic and lethal materials." That December, the annual meeting of the American Association for the Advancement of Science gave special attention to saving the environment and human civilization from "chemical invasion."[40]

As the year 1969 dawned, microbiologist and environmental activist René Dubos wondered which way the wind was blowing with present uses of science and technology. He asked, "Is this progress . . . or self destruction?" Dubos bemoaned the rising tide of environmental pollution and warned against contamination by "countless chemical substances." Cities, he wrote, could no longer benefit "from the cleansing effects of the winds for the simple reason that the wind itself is contaminated."[41] Dubos seemed to anticipate harrowing reports by toxicologists that numerous "chemical agents in our food, air, and water appear to be the largest cause of human cancers."[42] Dangerous chemicals at home, in nature, and at war appeared to be out of control.

Dubos concluded by citing the Skull Valley incident and criticizing the Army for using such chemicals before knowing how they moved through the environment. "The same criticism," Dubos cautioned, "can be leveled against society as a whole."[43] At least one US policymaker caught wind of these concerns and brought them before Congress and the American public.

THE SPECTER OF CHEMICAL DEATH

Early in 1969, NBC television aired a special on chemical and biological weapons that emphasized Seymour Hersh's revelations on lethal chemical weapons, reviewed the Skull Valley incident, and addressed the military's use of chemical agents in Vietnam. It so frightened New York Congressman Richard McCarthy that he delivered a speech in Congress lambasting the government's secrecy and contradictions regarding deadly chemical weapons, as well as their connection to tear gas. The US State Department, he explained, claimed to adhere to the Geneva Protocol of 1925, which banned the use in war of "asphyxiating, poisonous or other gases." even though the United States had never ratified the treaty. Yet, McCarthy railed, at that very moment, the United States was "engaging in chemical warfare in Vietnam by using tear gas to aid in the killing of the enemy!"[44] McCarthy's individual crusade gained little traction until his exposure of cross-country shipments of killer chemicals by the military tapped into popular concerns over national safety and environmental pollution.

That spring, McCarthy uncovered a secret Pentagon plan to dump into the ocean enormous amounts of deadly, leaking chemical warfare munitions only 125 miles offshore. The Department of Defense had labeled the plan Operation CHASE, shorthand for Cut-Holes-And-Sink-'Em. While it took less than 2.5 milligrams of VX gas to kill a person, the Pentagon planned to dump some 27,000 tons of it in one-ton containers. Even more frightening, the ocean dumping would occur only after shipping the chemicals by railway across two-thirds of the nation, from a chemical weapons facility near Denver to a port just twenty miles outside New York City.[45]

News that Operation CHASE would bring killer chemical weapons through several major population centers inspired protest at home and abroad. American media described the plan as "lamentably retarded," and labeled the railcars "Transcontinental 'Death Trains.'" The British *Guardian* denounced it as "morally untenable," while Irish Ambassador William Fay telephoned Congress members to convey fears of poisoning Ireland's fish and shores.[46] The outcries spurred congressional hearings that revealed the Army routinely dumped deadly toxins in the ocean, and had even scheduled one for the very next week. When the National Academy of Sciences ad hoc advisory committee recommended termination of ocean dumping, the Army scrapped the plan.[47]

That summer also saw nearly twenty American personnel rushed to the hospital after deadly VX nerve gas leaked at a secret US chemical weapons depot on Okinawa. The event occurred just before US Secretary of State William Rogers arrived in Tokyo to renew the US-Japan security treaty, which permitted US maintenance of military bases on Okinawa and elsewhere in Japan. Questions arose as to whether Washington had consulted Tokyo before storing the leaking lethal gases on Okinawa. Both Nixon and Rogers denied any knowledge of the policy. The security treaty was renewed, but not without ruffled feathers, and only upon mutual agreement that the chemical agents be removed from Okinawa.[48] Chemical weapons, it seemed, had earned as much public attention in 1969 as New York Jets quarterback Joe Namath or, perhaps more appropriately, as serial killer Charles Manson.

The press drew connections between lethal chemical weapons and tear gas in Vietnam. *US News and World Report*, for example, condemned Operation CHASE as "death on the move," just above an article that asked, "Antiriot Sprays: Are They Safe—Or Hazardous?"[49] In a separate article, Hersh cited tear gas use in Vietnam before explaining how "most of the World War I gas warfare deaths resulted from mustard gas, which was not introduced into combat until after both sides had tried tear gas."[50] The UN Secretary-General's annual report echoed these concerns, deriding both "the growing tendency to use some chemical agents for civilian riot control and a dangerous trend to accept their use in some form in conventional warfare."[51] Concerns about a slippery slope when using tear gas in war had finally gained public traction.

POLITICAL PRESSURES

Escalating chemical anxieties engendered political challenges, both internationally and at home, for the new Nixon administration. Internationally, several organizations in 1969, including the World Health Organization, the Stockholm International Peace Research Institute, and the US Senate, all released studies against chemical weapons proliferation. The UN Disarmament Committee's report on chemical and biological weapons called for their unconditional banishment from combat, arguing that any chemical weapon in war raised "a serious risk of escalation, both in the use of more dangerous weapons belonging to the same class, and of other weapons of mass destruction." UN Secretary-General U Thant described the proliferation of chemical and biological weapons as "far more serious" than nuclear weapons because they were "easily accessible to the poor countries."[52] America's use of Vietnam as a chemical laboratory not only threatened its moral leadership, it threatened global security.

These events compelled President Nixon's executive branch review of US policies regarding chemical weapons—the first such review in over fifteen years.[53] After analyzing advantages and disadvantages to its ratification, the review defined the Geneva Protocol of 1925 as an international "no first use" agreement. It did not ban research, production, or possession of "asphyxiating, poisonous, or other gases." It merely prohibited their use in war. The protocol's coverage of tear gas, however, remained ambiguous. While the protocol's English text forbade "other gases," the French version used the word "*similaires*," meaning only gases similar to asphyxiating or poisonous gases. Under typical conditions, tear gas neither asphyxiates nor poisons.

Debate arose between the US departments of defense and state regarding use of tear gases in war. The Department of Defense sought protocol ratification with an interpretation permitting use of nonlethal riot control agents, arguing, "the use of tear gases in conjunction with lethal weapons in Vietnam has made it easier to kill the enemy with fewer US casualties." The State Department, citing international concerns, noted how most nations "consider unrestricted use of tear gases to be (a) contrary to the Protocol, and (b) a loosening of the barriers against [chemical weapons] in general, and criticize the US use." The State Department worried that ratification of the Protocol without curtailing tear gas use in war would be "more difficult to 'sell' internationally . . . especially in the 'third world' where tear gas would be most applicable."[54] President Nixon had to make the final decision.

Nixon's chief national security advisor, Henry Kissinger, encouraged the President to capture "a net political advantage" by resubmitting the protocol for Senate ratification, while maintaining tear gas's benefits in Vietnam. Kissinger believed "either the Geneva Protocol doesn't mean anything or it doesn't apply to tear gas."[55] Two days before Thanksgiving in 1969, Nixon proclaimed he would resubmit the 1925 treaty to the Senate.[56] After initial celebration, both international and much domestic opinion soured when Nixon's tear gas and herbicide reservations to the protocol were made public.

Within weeks, the UN General Assembly overwhelmingly passed a resolution stating that international law, as embodied in the Geneva Protocol of 1925, prohibited the use of all chemical and biological agents in war, lethal and nonlethal. Only three nations opposed the resolution: the United States and Australia, both using tear gas and herbicides in Vietnam; and Portugal, then using tear gas against rebelling African colonies. The UN's decision delayed Nixon's submission of the protocol for eight months.[57] Advocates against tear gas in war used Nixon's delay to lobby Congress.

In Senate testimony, Harvard biologist and arms control activist Matthew Meselson challenged the humane and home arguments while opposing the tactical use argument with a slippery slope perspective. When Senator Milton Young emphasized the effective use of tear gas by domestic police, Meselson replied that military use of tear gas was "absolutely different." Soldiers, he explained, used tear gas in Vietnam not to save lives but "to increase enemy casualties." Meselson continued, "We should not allow ourselves to be confused that CS [tear gas] makes war more humane." Instead, tear gas use in war "set in motion a process" to prepare and defend against it, which, in turn, increased the enemy's chances to reply similarly or with deadly chemicals. Based on prior history, escalation to lethal chemical agents could result in American deaths. The small advantages afforded by tear gas in combat, Meselson concluded, was not worth "ushering in a new era of war." Ban all chemical weapons from war, he affirmed, including nonlethal tear gas and herbicides.[58]

Unconvinced, in 1970 President Nixon resubmitted the protocol to the Senate along with a memorandum granting exceptions for use of tear gas and defoliants in war. It required presidential approval of tear gas in future wars, but permitted ongoing use in Vietnam. The memorandum ignored the humane argument and supported the military's tactical use of tear gas to increase Vietcong deaths. It pressed the home argument, reiterating how tear gas in Vietnam was also "used domestically for riot control and law enforcement."[59]

By the end of the decade, although urban riots had significantly abated, civil protests in opposition to the Vietnam War had increased significantly, involving many more middle-class college students.[60] Some liberal commentators criticized the excessive use of tear gas against students, especially after the People's Park incident at the University of California, Berkeley in 1969.[61] In late spring 1970, campuses across the country erupted in opposition to Nixon's announced invasion of Cambodia. The horrific killing of students during campus protests at Kent State University in Ohio (see

FIG. 3.3. Antiwar demonstrations that turned lethal at Kent State University in Ohio on May 4, 1970. National Guard personnel walking toward Taylor Hall; tear gas has been fired. SOURCE: May 4 Collection, Box 28, Folder 705, Image 4-1-34. Courtesy of Kent State University Libraries, Special Collections and Archives.

fig. 3.3) and at Jackson State College in Mississippi in May underscored the limitations of current crowd-control techniques.[62] In this charged atmosphere, the Geneva Protocol debate again headed to the US Senate.

PASSING PROTOCOL

Before moving to a full Senate vote, the Geneva Protocol and the attached presidential memo had to pass through Senator Fulbright's Foreign Relations Committee. Early in 1971 the committee heard testimonies from Arthur Galston, a Yale University botanist who described herbicide use in Vietnam as ecocide,[63] and, again, from Meselson, who reiterated his slippery slope concerns, citing historical escalation from tear gas to lethal gas in World War I and most recently in Yemen. Meselson summarized his arguments succinctly: "1) The military value of riot gas to the United States is very low. 2) Our overriding security interest in the area of chemical and biological weapons is to prevent the proliferation and use of biological and lethal chemical weapons. 3) Our use of riot gas in war runs directly counter to this fundamental interest."[64]

After those hearings, Fulbright refused to submit the protocol for full Senate approval unless President Nixon reinterpreted the treaty "without restrictive understandings, or, if that is not possible at this time, . . . postpone further action on the Protocol until it is."[65] Nixon refused, awkwardly defending tear gas in war as "sound and basically humane."[66] Fulbright responded by freezing action on the protocol indefinitely. In 1973 major US military activities in Vietnam ended and with it the extensive use of tear gas in combat. That year, the Watergate crisis escalated and by August 1974 President Nixon resigned.

In an effort to accrue domestic and international political capital, President Gerald Ford's administration renewed efforts to ratify the protocol. In December 1974 Dr. Fred Ikle, director of the US Arms Control and Disarmament Agency, announced President Ford's willingness to renounce, as a matter of national policy, the first-use of tear gas in war except in limited, defensive circumstances. Those exceptions included riots at prisons or in other US military zones, in situations where enemies used civilians to shield attacks, in isolated rescue missions, and in rear echelon areas to protect convoys. Even then, according to this policy, the President "must approve in advance any use of riot-control agents and chemical herbicides in war." Within two days, on December 12, the protocol moved to the Senate floor and was approved unanimously four days later.[67]

On January 22, 1975, President Ford signed the instruments of ratification of the Geneva Protocol of 1925, formalizing US commitment against the first use of chemical and bacteriological weapons in war. The administration maintained that the protocol did not completely forbid use of tear gas or herbicides in war, but it did renounce their first use except, in Ford's words, "in a certain, very, very limited number of defense situations where lives can be saved."[68] On April 8, 1975, two days before the United States deposited the ratified protocol with the French government, Ford issued Executive Order 11850, which listed explicitly the exemptions for military tear gas use outlined originally by Dr. Ikle.[69] Ten years after American reintroduction of tear gas to international warfare, despite years of use without escalation to lethal chemicals, the slippery slope argument had triumphed, thereby establishing America's bifurcated policies on nonlethal chemical weapons. Today, tear gas still carries severe prohibitions as a wartime tool, yet domestic police forces and individual citizens maintain the liberty to apply it freely.

At the dawn of the Vietnam era, tear gas offered Americans a technological solution to social challenges at home and to the destructiveness of war abroad. By the early 1970s, however, tear gas transitioned from a nonviolent remedy to an international liability. The chemical did not change; social and political considerations on the contexts of its use did. Although tear gas remained an important domestic tool for controlling crowds nonviolently, a combination of chemical anxieties and social movements eventually generated opposition to nonlethal tear gases and herbicides as wartime weapons. A new environmentalism incited general disquiet over society's contamination by chemicals and spurred protest by concerned scientists and arms control advocates against the use of chemical weapons in Vietnam.[70] Additionally, a series of incidents at home and abroad revealed the acute dangers of lethal chemical weapons and escalated concerns of their eventual proliferation and use.

Collectively, these events earned domestic and international support for the slippery slope perspective, whereby use of any chemical weapons in war—including nonlethal ones—could incite a dangerous new era of warfare with unstable nations producing cheap killer chemicals for combat. After years of disagreement, Congress outlasted and overruled efforts by the Nixon administration and the Department of Defense to continue wartime use of nonlethal chemical weapons. In 1975, the United States finally ratified the Geneva Protocol of 1925, formally repealing the first use of chemical weapons in war, including tear gas. Paradoxically, American citizens and domestic police retained unrestricted use of the same nonlethal chemicals.

The paradox of American policies toward tear gas use domestically and abroad reflects a contradiction made clear in the Vietnam era about the authority of science and the production of scientific knowledge. This contradiction, as described by Stephen Bocking, is that "the authority of science, justified in terms of a capacity to generate facts true in every context, is continually recreated out of local materials, its particular form the product of specific circumstances."[71] Similarly, circumstances unique to the Vietnam era helped create America's paradoxical policies on tear gas, a scientific technology that generates different meanings when dispersed in different contexts.

The changing use of tear gas in the Vietnam era mirrors a simultaneous shift in the public's relationship to science and technology. Bifurcated domestic and military policies concerning use of tear gas illustrate an ambiguous complex of attitudes toward technology. In this paradox, broad populations considered tear gas a welcome and humane solution to diverse acts of domestic social unrest, while those same publics came to see tear gas in war as a threat to health and security. The local versus global discords of tear gas depict the challenge of regulating airborne chemicals across diverse political, physical, and social geographies.

NOTES

1. Sidell, "Riot Control"; Homan, "Tests."

2. Technological solutions to problems that, ultimately, require a combination of social, political, and technological overhaul remains a common theme in environmental histories of technology. See, for example, Dunn and Johnson, this volume.

3. Meselson, "Tear Gas in Vietnam," 17–19.

4. See Kilshaw, this volume, for more contemporary concerns about exposure to chemicals used in war.

5. The literature is extensive. See for example, Coleman, *History of Chemical Warfare*, and Zierler, *Invention of Ecocide*. For literature on nonlethal agents, see Davison, *"Non-Lethal" Weapons*.

6. Agar, "What Happened"; Moore, *Disrupting Science*, 190–214; Hughes, *American Genesis*, 443–72.

7. Gerald R. Ford, "Executive Order 11850: Renunciation of Certain Uses in War of Chemical Herbicides and Riot Control Agents, Apr. 8, 1975," The American Presidency Project, http://www.presidency.ucsb.edu/ws/?pid=59189#axzz1XUmqwNmV.

8. West, "History of Poison Gases," 415; Jones, "From Military to Civilian," 151–54, 165–68.

9. Norvel and Tuttle, "Views of a Negro"; "Urges 'Gassing'"; "Police to Use Tear Gas."

10. Jones, "American Chemists," 430–35; Slotten, "Humane Chemistry," 490–93; Meselson, "Tear Gas in Vietnam," 18; Edwards, *Korean War Almanac*, 494.

11. Sidell, "Riot Control Agents," 308, 310.

12. Frankel, "US Reveals"; SIPRI, *Problem*, 72; "Great Gas," 19–20.

13. "Great Gas," 20; "Gas and 'Marauding' Jet Raids," 12; "Asahi says US Loses Friends," 6; Hersh, *Chemical and Biological Warfare*, 170.

14. "Memorandum From the President's Special Assistant for National Security Affairs (Bundy) to President Johnson, Washington, Mar. 22, 1965," *Foreign Relations of the United States 1964–1968, Vol. II, Vietnam, Jan.–June 1965*, Washington, DC: Department of State, http://history.state.gov/historicaldocuments/frus1964-68v02/

d209; "Memorandum From the President's Special Assistant for National Security Affairs (Bundy) to President Johnson, Washington, Mar. 23, 1965," *Foreign Relations of the United States 1964–1968, Vol. II, Vietnam, Jan.–June 1965*, Washington, DC: Department of State, http://history.state.gov/historicaldocuments/frus1964–68vo2/ d210.

15. "Text of President Johnson," 639.

16. Savitz, "Gas and Guerrillas," 14; "Door Best Kept Closed," 5–6; "Gas: Nobody's Mistake," 4.

17. "New Backing," 6.

18. Pleasants, "Gas Warfare," 209, 212.

19. McCarthy, *Ultimate Folly*, 46.

20. Apple, "Marine Officer," 1; "Protest over the Use of Gas," 7.

21. Apple, "US Colonel Absolved," 1; "Tear Gas in Vietnam," 8; SIPRI, *Problem*, 188–89.

22. "How Gas Is Being Used," 8.

23. Margolis, "From Washington," 31.

24. "US Explains New Tactic," 2; Savitz, "Gas and Guerrillas," 13–14.

25. Bloch et al., "Scientists Speak Out," 39–40.

26. Utgoff, *Challenge*, 73.

27. Bart, "New Negro Riots," 1; "City Riots," 175–180; "Fire This Time"; "As Rioting Spread," 26–28; "Race Troubles," 28–30. See also Isserman and Kazin, *America Divided*.

28. Zacks, "Coming: Fantastic Devices," 62–64; "Police: Disabling Without Killing," 50–52; Sagalyn and Coates, "Wanted," 6, 12–16; "Humane Riot Control," 165–66; "How to Stop Riots," 132–35; "New Nonlethal Weapons," 76–77; Swearengen, *Tear Gas Munitions*; Fennessy et al., "Beyond Vietnam."

29. Girardin, "Speaking Out."

30. "Search for Peace in Asia."

31. US FBI, *Prevention*, 13, 65, 91, 83.

32. *Army Digest* quoted in McCarthy, *Ultimate Folly*, 50.

33. SIPRI, *Problem*, 190; McCarthy, *Ultimate Folly*, 49; US Department of Justice, Civil Rights Division, "Omnibus Crime Control and Safe Streets Act of 1968, 42 USC. § 3789d," accessed Apr. 12, 2012, http://www.usdoj.gov/crt/about/spl/42usc3789d .php.

34. Carson, *Silent Spring*; Guha, *Environmentalism*, 69–73.

35. "CW in Vietnam," 7; "Breath of Death," 18–19; "Chemical and Biological Weapons," 1398–1403; "Congressional Fight Brews," 2385.

36. "45 Yemenis Reported Killed," 9; "In New Detail," 9; SIPRI, *Problem*, 160.

37. "It Was the Sheep," 589.

38. Boffey, "6000 Sheep Stricken," 1442; "Sheep and the Army," 104; Boffey, "Sheep Die," 327–329. See also McCarthy, *Ultimate Folly*, 109–11.

39. Hersh, *Chemical and Biological Warfare*; Hersh, "Just a Drop Can Kill"; Hersh, "Our Chemical War"; Hersh, "Gas and Germ Warfare"; Hersh, "Chemical and Biological Weapons."

40. Bengelsdorf, "War Against the Land," L1; Neilands, "More on Forest Defoliation," 965; "Saving Man's Environment," 30.

41. Dubos, "Is This Progress," 142.

42. Kotulak, "Links Cancer to Chemicals," C25; "Tests With Mice," A21.

43. Dubos, "Is This Progress," 142.

44. Finney, "Pentagon Scored," 2; Hersh, "Germs and Gas," 13–16.

45. "Poisonous Gas Disposal," 809–10; McCarthy, *Ultimate Folly*, 99–109; Hersh, "Germs and Gas," 14.

46. "How Not to Do It," 653; "Transcontinental 'Death Trains,'" 817; McCarthy, *Ultimate Folly*, 106.

47. US NAS, *Report of the Disposal Hazards*.

48. Sheehan, "US Said to Keep Nerve Gas," 1; "Okinawa Agreement," 2386–87; McCarthy, *Ultimate Folly*, 129–32; Utgoff, *Challenge*, 137.

49. "Poison Gas"; "Antiriot Sprays," 10.

50. Hersh, "Chemical and Biological Weapons," 82.

51. UN report quoted in US Senate Special Subcommittee, *Chemical and Biological Weapons*, 18.

52. Boffey, "CBW," 1378; "Conclusion of the UN Report," 10; Nelson, "Arms Control," 1108.

53. "President's Talking Points."

54. "Analytical Summary Prepared by the National Security Council Staff, Washington, undated [Oct. 1969]," *Foreign Relations of the United States 1969–1976, Volume E-2, Documents on Arms Control, 1969–1972*, Washington, DC: Department of State, http://history.state.gov/historicaldocuments/frus1969–76ve02/d152; "Germ Warfare," 2386.

55. "Minutes of National Security Council Review Group Meeting, Washington, Oct. 30, 1969, 2:25–3:55 p.m.," *Foreign Relations of the United States 1969–1976, Volume E-2, Documents on Arms Control, 1969–1972*, Washington, DC: Department of State, http://history.state.gov/historicaldocuments/frus1969–76ve02/d155.

56. "President's Talking Points"; Naughton, "Nixon Renounces," 1; Shuster, "Capitals in Western Europe," 17.

57. United Nations General Assembly, "Resolution No. A/RES/2603(XXIV): Question of Chemical and Bacteriological (Biological) Weapons, December 16,

1969," http://www.un.org/Depts/dhl/resguide/r24.htm; "Germ Warfare," 2385; Utgoff, *Challenge*, 101.

58. Meselson, "Tear Gas in Vietnam," 17–19.

59. "National Security Decision Memorandum 78, Washington, Aug. 11, 1970," *Foreign Relations of the United States 1969–1976, Volume E-2, Documents on Arms Control, 1969–1972*, Washington, DC: Department of State, http://history.state.gov/historicaldocuments/frus1969–76veo2/d202.

60. "End to Big Riots," 42–44; "How '69 Riot Season Compares," 16; DeBenedetti and Chatfield, *American Ordeal*, 203–311.

61. Berry et al., "Berkeley Park," 784–88; Lando, "It's a Gas!," 17–18.

62. President's Commission, *Report*.

63. Zierler, *Invention of Ecocide*, 15.

64. Meselson, "Gas Warfare," 33–37.

65. "Memorandum From the President's Assistant for National Security Affairs (Kissinger) to the Secretary of State Rogers and Secretary of Defense Laird, Washington, June 28, 1971," *Foreign Relations of the United States 1964–1968, Vol. II, Vietnam, Jan.–June 1965*, Washington, DC: Department of State, http://history.state.gov/historicaldocuments/frus1969–76veo2/d232.

66. "Memorandum From the Executive Secretary of the Department of State (Eliot) to the President's Assistant for National Security Affairs (Kissinger), Washington, Mar. 31, 1971," *Foreign Relations of the United States 1969–1976, Vol. E-2, Documents on Arms Control, 1969–1972*, Washington, DC: Department of State, http://history.state.gov/historicaldocuments/frus1969–76veo2/d223.

67. Federation of American Scientists, "Geneva Protocol," accessed Apr. 12, 2012, http://www.fas.org/nuke/control/geneva/intro.htm.

68. Gerald Ford, "Remarks upon Signing Instruments of Ratification of the Geneva Protocol of 1925 and the Biological Weapons Convention, Jan. 22, 1975," The American Presidency Project, http://www.presidency.ucsb.edu/ws/?pid=5038#axzz1XUmqwNmV.

69. Ford, "Executive Order 11850."

70. Zierler, *Invention of Ecocide*.

71. Bocking, *Nature's Experts*, 25.

REFERENCES

Agar, Jon. "What Happened in the Sixties?" *British Journal for the History of Science* 41, no. 4 (2008): 567–600.

"Antiriot Sprays: Are They Safe—or Hazardous?" *US News and World Report*, June 2, 1969.

Apple, R. W., Jr. "Marine Officer Uses Tear Gas in Vietnam, Setting Off Inquiry." *New York Times*, Sept. 8, 1965.

———. "US Colonel Absolved in Vietnam Tear Gas Incident." *New York Times*, Sept. 25, 1965.

"As Rioting Spreads . . . The Search for Answers." *US News and World Report*, Aug. 14, 1967.

"Asahi says US Loses Friends." *New York Times*, Mar. 24, 1965.

Bart, Peter. "New Negro Riots Erupt on Coast; Three Reported Shot." *New York Times*, Aug. 13, 1965.

Bengelsdorf, Irving S. "War Against the Land: The Land—New Victim of Viet War." *Los Angeles Times*, June 16, 1968.

Berry, Frederick, Thomas Brooks, and Eugene Commins. "The Berkeley Park: Terror in a Teapot." *Nation*, June 23, 1969.

Bloch, Felix, et al. "Scientists Speak Out on CB Weapons." *Bulletin of the Atomic Scientists* (Nov. 1966): 39–40.

Bocking, Stephen. *Nature's Experts: Science, Politics, and the Environment*. New Brunswick, NJ: Rutgers University Press, 2006.

Boffey, Phillip M. "6000 Sheep Stricken near CBW Center." *Science* 154 (1968): 1442.

———. "CBW: Pressures for Control Build in Congress, International Groups." *Science* 164 (1969): 1378.

———. "Sheep Die near Nerve Gas Tests." *Science News*, Apr. 6, 1968.

"A Breath of Death: The Dilemma of Chemical and Biological Warfare." *Senior Scholastic*, Feb. 7, 1969, 3–6, 18–19.

Carson, Rachel. *Silent Spring*. Greenwich, CT: Fawcett Publications, 1962.

"Chemical and Biological Weapons Stir Controversy." *Congressional Quarterly Weekly Report* 27, no. 31 (Aug. 1, 1969): 1398–1403.

"City Riots—A Worldwide Plague." *Harper's Magazine*, Nov. 1965. Reprinted in *Reader's Digest*, Feb. 1966.

Coleman, Kim. *A History of Chemical Warfare*. New York: Palgrave Macmillan, 2005.

"Conclusion of the UN Report on Chemical and Biological Weapons." *New York Times*, July 3, 1969.

"Congressional Fight Brews (Over CS-2 Tear Gas and Defoliants)." *Congressional Quarterly Weekly Report* 27, no. 48 (Nov. 28, 1969): 2385.

"CW in Vietnam? Defoliants and Tear Gas." *Senior Scholastic*, Feb. 7, 1969.

Davison, Neil. *"Non-Lethal" Weapons*. Basingstoke, UK: Palgrave Macmillan, 2009.

DeBenedetti, Charles, and Charles Chatfield. *An American Ordeal: The Antiwar Movement of the Vietnam Era*. Syracuse, NY: Syracuse University Press, 1990.

"A Door Best Kept Closed." *New Republic*, Apr. 3, 1965.

Dubos, René. "Is This Progress . . . or Self-Destruction?" *New York Times*, Jan. 6, 1969.

Edwards, Paul M. *Korean War Almanac*. New York: Infobase Publishing, 2006.

"End to Big Riots? Findings of a Police Survey." *US News and World Report*, June 2, 1969.

Fennessy, E. F., Jr., J. A. Russo Jr., and R. H. Ellis. "Beyond Vietnam: What Has Science to Say to Man? Humane Policing." *Saturday Review*, July 1, 1967.

Finney, John W. "Pentagon Scored on Chemical War." *New York Times*, Apr. 22, 1969.

"The Fire This Time." *TIME*, Aug. 4, 1967.

"45 Yemenis Reported Killed (By Gas)." *New York Times*, July 7, 1967.

Frankel, Max. "US Reveals Use of Nonlethal Gas Against Vietcong." *New York Times*, Mar. 24, 1965.

"Gas: Nobody's Mistake but L.B.J.'s." *LIFE*, Apr. 2, 1965.

"Gas and 'Marauding' Jet Raids in Vietnam." *Times* (London), Mar. 23, 1965.

"Germ Warfare." *Congressional Quarterly Weekly Report* 27, no. 48 (Nov. 28, 1969): 2386.

Girardin, Ray. "Speaking Out; After the Riots: Force Won't Settle Anything." *Saturday Evening Post*, Sept. 23, 1967.

"The Great Gas Flap." *TIME*, Apr. 2, 1965.

Guha, Ramchandra. *Environmentalism: A Global History*. New York: Addison Wesley Longman, 2000.

Hersh, Seymour M. *Chemical and Biological Warfare: America's Hidden Arsenal*. New York: Bobbs-Merrill Co., 1968.

———. "Chemical and Biological Weapons: The Secret Arsenal." *New York Times Magazine*, Aug. 25, 1968.

———. "Gas and Germ Warfare: The Controversy over Campus Research Contracts." *New Republic*, July 1, 1967.

———. "Just a Drop Can Kill: Secret Work on Gas and Germ Warfare." *New Republic*, May 6, 1967.

———. "Our Chemical War." *New York Review of Books*, May 9, 1968.

Homan, Richard. "Tests Show Riot Gas Hurts Skin." *Washington Post*, June 30, 1969.

"How '69 Riot Season Compares with 1968." *US News and World Report*, Sept. 15, 1969.

"How Gas Is Being Used in Vietnam." *US News and World Report*, Jan. 31, 1966.

"How Not to Do It." *Nation*, May 26, 1969.

"How to Stop Riots." *Reader's Digest*, Oct., 1967.

Hughes, Thomas P. *American Genesis: A Century of Invention and Technological Enthusiasm, 1870–1970*. Chicago: University of Chicago Press, 1989, 2004.

"Humane Riot Control." *The American City: Magazine of Municipal Management and Engineering*, Sept., 1967.

"In New Detail—Nasser's Gas War." *US News and World Report*, July 10, 1967.

Isserman, Maurice, and Michael Kazin. *America Divided: The Civil War of the 1960s*. New York: Oxford University Press, 2000.

"It Was the Sheep This Time." *Nation*, May 6, 1968.

Jones, Daniel P. "American Chemists and the Geneva Protocol." *Isis* 71 (1980): 426–40.

———. "From Military to Civilian Technology: The Introduction of Tear Gas for Civil Riot Control." *Technology and Culture* 19 (1978): 151–68.

Kotulak, Ronald. "Links Cancer to Chemicals We Live With." *Chicago Tribune*, Mar. 30, 1969.

Lando, Barry. "It's a Gas!" *New Republic*, June 28, 1969.

Margolis, Howard. "From Washington: Notes on Gas and Disarmament." *Bulletin of the Atomic Scientists* (Nov. 1965): 30–32.

McCarthy, Richard D. *The Ultimate Folly: War by Pestilence, Asphyxiation, and Defoliation*. New York: Alfred A. Knopf, 1969.

Meselson, Matthew. "Gas Warfare and the Geneva Protocol of 1925." *Bulletin of the Atomic Scientists* (Feb. 1972): 33–37.

———. "Tear Gas in Vietnam and the Return of Poison Gas." *Bulletin of the Atomic Scientists* (Mar. 1971): 17–20.

Moore, Kelly. *Disrupting Science: Social Movements, American Scientists, and the Politics of the Military, 1945–1975*. Princeton, NJ: Princeton University Press, 2008.

Naughton, James M. "Nixon Renounces Germ Weapons, Orders Destruction of Stocks; Restricts Use of Chemical Arms." *New York Times*, Nov. 26, 1969.

Neilands, J. B. "More on Forest Defoliation." *Science* 161 (1968): 965.

Nelson, Bryce. "Arms Control: Demand for Decisions." *Science* 162 (1968): 1108.

"New Backing for the Use of Nausea Gas." *US News and World Report*, Apr. 12, 1965.

"New Nonlethal Weapons Give Police Potent Alternative to Club and Gun." *LIFE*, Nov. 24, 1967.

Norvel, Stanley B., and William M. Tuttle Jr. "Views of a Negro during the 'Red Summer' of 1919." *Journal of Negro History* 51 (1966): 209–18.

"Okinawa Agreement." *Congressional Quarterly Weekly Report*, 27, no. 48 (Nov. 1969): 2386–87.

Pleasants, Julian. "Gas Warfare: Is It Justifiable as Minimum Force?" *Commonweal* 7 (May 1965): 209–12.

"Poisonous Gas Disposal." *Congressional Quarterly Weekly Report* 27, no. 21 (May 23, 1969): 809–10.

"Poison Gas—27,000-Ton Headache." *US News and World Report*, June 2, 1969.

President's Commission on Campus Unrest. *The Report of the President's Commission on Campus Unrest.* Washington, DC: US Government Printing Office, 1970.

"The President's Talking Points: Congressional Leadership Meeting, Top Secret, Talking Points, Nov. 25, 1969." *Digital National Security Archive.* ProQuest (TE00061).

"Protest over the Use of Gas." *Times* (London), Sept. 10, 1966.

"Police: Disabling Without Killing." *TIME*, May 5, 1967.

"Police to Use Tear Gas." *New York Times*, Oct. 27, 1921.

"Race Troubles: Record of 109 Cities." *US News and World Report*, Aug. 14, 1967.

Sagalyn, Arnold, and Joseph Coates. "Wanted: Weapons that Do Not Kill." *New York Times Magazine*, Sept. 17, 1967.

"Saving Man's Environment." *New York Times*, Dec. 30, 1968.

Savitz, David. "Gas and Guerrillas—A Word of Caution." *New Republic*, Mar. 19, 1966.

"The Search for Peace in Asia, the Specter of Violence at Home" (Cover). *TIME*, Apr. 12, 1968.

Sheehan, Neil. "US Said to Keep Nerve Gas Abroad at Major Bases: Report of Okinawa Accident Sets Off Furor in Japan." *New York Times*, July 19, 1969.

"Sheep and the Army." *TIME*, Apr. 5, 1968.

Shuster, Alvin. "Capitals in Western Europe Welcome Nixon's Move." *New York Times*, Nov. 26, 1969.

Sidell, Frederick R. "Riot Control Agents." In *Medical Aspects of Chemical and Biological Warfare*, edited by Frederick R. Sidell, Ernest T. Kakafuji, and David R. Franz, 307–24. Washington, DC: Borden Institute, Walter Reed Medical Center, 1997.

SIPRI (Stockholm International Peace Research Institute). *The Problem of Chemical and Biological Warfare: Vol. I: The Rise of CB Weapons.* Stockholm, Sweden: SIPRI, 1971.

Slotten, Hugh R. "Humane Chemistry or Scientific Barbarism? American Responses to World War I Poison Gas, 1915–1930." *Journal of American History* 77, no. 2 (Sep. 1990): 476–498.

Swearengen, Thomas F. *Tear Gas Munitions: An Analysis of Commercial Riot Gas Guns, Tear Gas Projectiles, Grenades, Small Arms Ammunition, and Related Tear Gas Devices.* Springfield, IL: Charles C. Thomas, Publisher, 1966.

"Tear Gas in Vietnam—Freer Use Ahead?" *US News and World Report*, Oct. 4, 1965.

"Tests With Mice: Common Chemicals, Cancer Link Studied." *Los Angeles Times*, Apr. 17, 1969.

"Text of President Johnson's April 1 News Conference." *Congressional Quarterly Weekly Report* 23, no. 15 (Apr. 9, 1965): 638–39.

"Transcontinental 'Death Trains.'" *Christian Century*, June 11, 1969.

"US Explains New Tactic." *New York Times*, Feb. 22, 1966.

US Federal Bureau of Investigation. *Prevention and Control of Mobs and Riots.* Washington, DC: United States Department of Justice, 1965, 1967.

US National Academy of Sciences (NAS), Ad Hoc Advisory Committee. *Report of the Disposal Hazards of Certain Chemical Warfare Agents and Munitions.* Washington, DC: National Academy Press, 1969.

US Senate, Special Subcommittee on the National Science Foundation of the Committee on Labor and Public Welfare. *Chemical and Biological Weapons: Some Possible Approaches for Lessening the Threat and Danger.* Washington, DC: US Government Printing Office, 1969.

"Urges 'Gassing' by Police: Baltimore Professor Advocates Use of Bombs in Attacks." *New York Times*, Mar. 22, 1920.

Utgoff, Victor A. *The Challenge of Chemical Weapons: An American Perspective.* London: Macmillan, 1990.

West, Clarence J. "The History of Poison Gases." *Science* 49 (1919): 412–17.

Zacks, Robert. "Coming: Fantastic Devices to End Riots." *Nation's Business*, July 1966.

Zierler, David. *The Invention of Ecocide: Agent Orange, and the Scientists Who Changed the Way We Think about the Environment.* Athens: University of Georgia Press, 2011.

4

TOXIC SOLDIERS
Chemicals and the Bodies of Gulf War Syndrome Sufferers

Susie Kilshaw

Sufferers of Gulf War syndrome worry about the risky and dangerous atmosphere to which they were exposed during the war and the lingering consequences of that exposure. The majority of veterans has a broad understanding of and anxiety about the role of chemicals in their illness, not only through chemical weapons, chemical warfare, and past hazards specific to their Gulf War experience but through enhanced sensitivity to ever-present (and ever-changing) toxins in their local environments. The exposures of the Gulf War in 1991 have left these soldiers vulnerable and thus they are left at risk of potentially hazardous chemicals they encounter in their daily lives. Their relationships with their bodies and their interactions with the environment around them are irrevocably altered by their experiences in the war. Veterans' anxiety about chemicals reflects a wider cultural anxiety in the United Kingdom and United States surrounding chemicals and toxins and implicates wider social, political, and economic dimensions in understandings about their illness. In this chapter I explore the way cultural imaginings feed back into beliefs and experiences of illness. My main research interest lies with the way some concerns become the focus of public attention and others are ignored. As a medical anthropologist I focus on issues of health and illness and how concerns about the atmosphere are very much connected to widespread cultural health anxieties and anxieties about the impact of humans on the environment.

THE PROJECT

On August 2, 1990, Iraq, led by Saddam Hussein, invaded and annexed the neighboring Persian Gulf state of Kuwait, thereby gaining control over some 20 percent of the world's oil reserves as well as access to ports on the Gulf. The United Nations Security Council unanimously denounced the invasion, demanded an immediate withdrawal of all forces, and imposed a worldwide ban on trade with Iraq. Within a week, some 230,000 US troops were deployed to Saudi Arabia under Operation Desert Shield to deter further expansion in the region, while Iraq deployed about 300,000 troops to Kuwait. The United States responded by sending an additional 200,000 troops, building a UN-authorized coalition of thirty-four nations, and sponsoring a UN resolution that established a deadline of January 15, 1991, for Iraq's complete withdrawal from the region. The United Kingdom deployed 53,462 personnel in an operation known as Operation Granby.

With the passing of the deadline, a massive allied aerial bombardment, code name Desert Storm, began on January 17. This was followed by an allied ground assault on February 23 that liberated Kuwait and devastated the Iraqi army one hundred hours after it had begun. The official ceasefire was accepted and signed on April 6. Aerial and ground combat were confined to Iraq, Kuwait, and areas on the border of Saudi Arabia. The war, characterized by its media presence and high-tech equipment, was often referred to as a "smart, clean, war," yet images of the conflict depicted scenes of burned-out, absolutely devastated Iraqi military columns and the dirty smoke from over 500 of Kuwait's burning oil wells clouding the horizon. A total of 482 Coalition troops died: 190 caused by enemy action (114 US, 38 UK), and the rest were through friendly fire (35 US, 9 UK) and accidents.[1]

In 1992 reports began to surface in the United States about unexplained health problems occurring among returning Gulf War veterans. The illness later appeared in the United Kingdom with the first reported cases in 1993. Many veterans expressed their conviction that they suffered from a unique and new disorder. Their explanations included exposure to chemical warfare agents, vaccinations, nerve agent pretreatment sets (NAPS) tablets, toxic fumes from burning oil wells, depleted uranium, and organophosphate insecticides, yet investigations have produced no compelling evidence of an organic syndrome. A great deal of investigation has been done into

Gulf War syndrome (GWS), but the illness remains perplexing. My research goes beyond the previous focus on medical and epidemiological research to understand GWS from an anthropological point of view.

Between September 2001 and November 2002, I conducted ethnographic fieldwork in the UK GWS community, interviewing veterans and their advocates as well as clinicians and researchers. The main focus was on the sufferers themselves and what they had to say about their illness. I conducted a total of 93 interviews, 67 of which were with UK Gulf War veterans, the majority of whom believed themselves to be ill with the condition. In addition to formal interviews, I maintained contact with informants, allowing for more informal discussions and observations. I also collected data from media files and other relevant documentation, including transcribed interviews with sufferers, family members, advocates, practitioners, scientists, and researchers. I also noted observations of interactions among those involved in the arena of GWS, including clinical encounters, formal meetings, and using informal discussions with researchers and scientists to explore further the biomedical and mainstream discourse surrounding GWS and the ways in which this was negotiated by sufferers. Unlike other research into this illness, my work focuses on sufferers' own accounts to understand better the way GWS is perceived by those it affects. I have presented some of this material elsewhere, but in this chapter I focus on fear of the environment and more specifically, a fear of chemicals in the atmosphere.[2]

CHEMICAL WEAPONS, CHEMICAL AIR

Gulf War veterans make sense of their experiences in particular ways. They understand their suffering to be caused by various exposures: depleted uranium, chemical weapons, organophosphates (chemical weapons, pesticides, and insect repellents), and oil fires, in addition to vaccinations and other preventative measures. Various theories encompassing different exposures sit side by side, sometimes overlap, and sometimes contradict each other.[3] At times a particular theory will come to the fore and at other times it will be forgotten. But what remains is the focus on chemicals, toxins, and the notion that there was something unnatural about the exposures encountered. Sufferers saw the environment as risky, unnatural, and dangerous—and this goes beyond the immediate war environment to include the atmosphere in a more general sense. Veterans often would describe the Gulf environment as strange and dangerous. They would describe an environment thick with smoke and the sky turning black from the oil well fires. "Day turned to night," they would say. Dust was ever-present, and smells were odd and unusual. Of course, there were also fears about chemicals lingering in the air.

War is always risky and there are never shortages of possible hazards to health in the environment of war. Furthermore, as Ana Carden-Coyne and others have shown, chemicals have been a concern of soldiers and their leaders throughout history.[4] The Gulf War, however, introduced novel and unexpected hazards including the smoke from the oil fires that blackened the sky and polluted the air. There were also "other hazards that were more recent additions to the ways in which mankind has harnessed technology to deadly ends."[5] In 1991 Saddam Hussein possessed large stocks of chemical and biological weapons, and the threat of their use was very real, for he had used them against Kurdish civilians.

In my interviews, soldiers expressed concerns about chemical weapons through reporting of the frequency of the alarms sounding, which were used to detect chemical weapons. It was the terror or the threat rather than the reality of them that veterans remember. Participants recall frequently hearing the alarms, which they regularly ignored, something that many use to support their theory of chemical poisoning. The *Riegle Report* of 1994 on the health of Gulf War veterans (authored by US Senator Donald W. Riegle Jr.) reported that chemical alarms went off 18,000 times during the Gulf War, suggesting that Coalition forces were continually exposed to low levels of chemical agents throughout the war. However, it was later suggested that the alarms were too sensitive and reacted to things like jet fuel.

Still, it is impossible to ignore the capacity of chemical weapons to terrorize. Recent studies of US veterans exposed to the threat of chemical weapons have shown that both symptoms and the memory of alerts in war zones are important in establishing and maintaining beliefs about being poisoned. In 2006, Brewer and colleagues reported that 64 percent of a sample of 335 US veterans of the Gulf War believed that they had been subjected to chemical weapons compared with 6 percent of 269 service controls who had not deployed to the conflict.[6] Veterans would often go into great details about having to suit up: if an alarm were sounded they would have to don their cumbersome and heavy nuclear, biological, chemical (NBC) suits quickly. Many expressed anxiety about this process and explained how difficult it was to do so and how the heat made it an uncomfortable exercise.

Some argued that Iraqi forces used sarin, with certain advocates claiming that the results of studies suggest that this happened on day four of the ground war.[7] However, most attention has been given to the possibility that a cause of ill health was the accidental discharge of sarin nerve agents that followed the postwar demolition of chemically armed rockets at the Khamisayah arms dump. The US military initially denied that the soldiers had been exposed to chemical weapons, but later stated that agents were released as the result of the destruction of the Khamisayah site. At a federal investigator's meeting in 2001, a prominent advocate of this theory suggested that the Khamisayah episode was a CIA smokescreen to cover up the real facts about deliberate use of sarin by the Iraqis.[8] When it was first reported, 400 veterans were said to be at risk. This number then rose steadily to 21,000. The fact that it took so long for the American government to disclose the event added to the accusation of a cover-up.

One prominent advocate and scientist had great influence with the veterans: he spoke often about various exposures and chemicals: "Oops, we weren't exposed to sarin gas, says the MoD [Ministry of Defence]. The Americans have now admitted that they were. We are still playing silly buggers at this, sorry about the language, but I just get so cross . . . 'Oh, there's no evidence that we were exposed to sarin gas.' What about all the alarms that went off? 'Oh they didn't work.' Why did you buy them, then? Why did you claim there was no exposure if the alarms went off anyway?"[9] Here he is referring to sarin being released when the Khamisayah rockets were destroyed. When I was interviewing veterans there was discussion about the plume created by the destruction of the Khamisayah arms dump. They would show me maps, which detailed the plume and where they were in relation to it. Soldiers who were downwind of or in the close proximity of the plume thought they had been affected. However, experts often revised the diagram, and participants would become agitated if changes to the plume location meant they had fallen out of its boundaries. One of my key informants with whom I met on several occasions often focused on the Khamisayah plume and his exposure to chemical weapons, but later he dismissed the role of chemical weapons in favor of depleted uranium (DU). He said this was, in part, due to the increased attention DU was getting in the media and its links to the complaints of veterans who fought in Bosnia and Kosovo, but also because he had come to accept that if chemical weapons had been used people would have died instantly. Indeed, I witnessed this deemphasizing of the role of chemical weapons throughout my fieldwork.

Extensive investigation and review by several expert panels have determined that no evidence exists that chemical warfare nerve agents were used during the Gulf War and it is unlikely that exposure to chemical warfare agents caused GWS.[10] The suggestion of exposure linked to the Khamisayah incident continues to be unconvincing, not least because of the absence of any contemporary evidence of adverse effects of exposure.[11] Furthermore, such an explanation simply does not incorporate the entirety of the GWS case. Yet the psychological impact of a perceived chemical warfare attack can still have immediate and long-term health consequences. The deployment- or war-related health impact from experiences of the Gulf War, including the perceived exposure to chemical warfare agents, should be considered as an important cause of morbidity among Gulf War veterans.[12] The effect of such exposures is unknown and nebulous, which provides the sufferer space to develop theories that incorporate and attempt to make sense of their experience of ill health. The threat of chemical weapons provided an anchor and a focus for beliefs about being poisoned. The theories then stem outwards and include other toxins. Interestingly, chemical weapons then take a back seat, although they retain a vague presence in illness narratives. But what remains is an overwhelming belief about being poisoned and being surrounded by harmful toxins. One veteran asked, "Are we carrying around these chemicals in our body and you know, are they still there?"

What remains is a much wider concern about the air and the environment and the notion of toxicity. Josh, a "well" veteran who was concerned about GWS but did not feel he had the condition, spoke about his feelings of toxicity: "I definitely have the feeling that I've been contaminated. Toxic war. Ties into the contamination thing. I don't feel I'm contaminating you by sitting here, but contaminating relationships . . . that their futures may not be as bright because of what I've been party to, what I've done, inhaled, injected with. . . . to do with reproduction, etc. It clouds my future and progeny's future. We've been exposed to so many chemicals. The chances of birth defects are vastly increased. It may not be noticeable in great scheme of things, but for my family it's bloody important." He continued:

> The government is losing hold on our bodies. Inhabit our bodies, but didn't have a say. We are meant to be in charge of our own bodies. We weren't in control of our own bodies. . . . Every chemical has an impact on our cells either the way they reproduce or the nucleus . . . will change in your body. . . . The chemical we had: we are going to have defects. Whether or not they are passed down the line but probably are. Everything we taste, smell, and touch might have effect on progeny because they are chemicals that we have put into our bodies. Impact of chemicals will only come out in hundreds of years.

Indeed, Josh is representative of many of the well veterans I met who, despite not having the condition, were concerned that Gulf War exposures had still affected them in some way. There was the anxiety that either the toxins had affected them in some hidden or unknown way, and that the impact would only be felt in the future. They felt as though no one was safe. Those who are not ill may merely be not ill yet. Concerns seemed primarily to center on issues having to do with reproduction: infertility and/or birth defects.[13]

Although my work focused on anxieties about Gulf War chemicals and the environment as they manifested in the United Kingdom, concerns about Gulf War chemicals had an impact much closer to their source—something about which I am now becoming aware due to my present research in the Middle East. Over the past three years I have been conducting fieldwork in Qatar, a small country that comprises a peninsula surrounded on most sides by the Arabian Gulf.[14] Despite embarking on what I thought was an entirely new project and subject matter, I have found that my interviews have brought me back to where I began: the Gulf War. One project investigates genetic knowledge among Qataris, particularly in relation to cousin marriage and disability—a subject distant from the concerns of UK Gulf War veterans. However, during these interviews I was struck by how many Qataris talk about the Gulf War of 1991 and their suspicions about the chemicals released during that time. Some have suggested that they suspect that there are increased rates of cancers and birth defects as a result of the toxins of the war.

CHEMICAL IMAGININGS: THE LEGACY OF PREVIOUS WARS

As an anthropologist I am interested in the impact of culture on beliefs and behaviors. Cultural imaginings influence reactions to chemical war and chemicals more generally, which are linked to the legacy of previous wars. Research on the experiences of World War I veterans has shown that the terror inspired by chemical weapons served to maintain memories of being gassed.[15] Unlike a bullet or piece of shrapnel, which could lodge in the body and be removed surgically, gas was systemic and had no definite physical limits. During WWI, gas was one of the most feared weapons. It inspired negative emotions out of all proportion to its ability to kill or wound.[16] Nearly a century later, chemical weapons have retained their capacity to frighten. Understanding the long-lasting effects of chemical weapons on the bodies and minds of WWI soldiers might assist in understanding the otherwise baffling persistence of ill health known as GWS.[17] The conviction of having been gassed had long-term negative effects on a person's beliefs about illness and perceptions of health and well-being. WWI veterans linked their accounts of work difficulties and chronic symptoms to enduring beliefs about being gassed.[18] Similarly, among Gulf veterans, experiences of illness are significantly associated with the belief that the Iraqis used chemical weapons. The British veterans I met sometimes pointed to the checkered history of organophosphates and their use in WWI.

Not only were chemical weapons a source of terror in WWI but the anti-gas measures used against such an attack were themselves disconcerting or a source of discomfort. Similarly, in the Gulf War measures to protect against the threat of chemical and biological weapons often generated anxiety and affected long-term health. Soldiers' bodies were not tough enough; they had to be fortified and extended and doing so emphasized the vulnerability and inadequacies of the human body. Soldiers were given a series of vaccinations and were ordered to take pyridostigmine bromide tablets to protect against organophosphate exposures. Through these prophylactic measures, soldiers' bodies were altered and strengthened to protect them against the threat of chemical weapons: internal body boundaries were bolstered.[19] External body boundaries were fortified by the use of NBC suits and masks. With such a great deal of focus on fortifying the body, one must ask whether bodies are experienced as more vulnerable once they are stripped of this armor, once they return home. Furthermore, the vaccinations and medications given to them to protect them were, in the end, seen as harmful. Rather than thinking of these prophylactic measures as bolstering and protecting, they were seen to weaken the body and the immune system. Furthermore, sufferers suggested that internal body boundaries became permeable and weak. The preventative measures take center stage in theories of causation for the illness.

The veterans' relationship with the environment is altered: their body boundaries are porous and their immune systems damaged, leaving the body open and vulnerable to the environment. The immune system has become the means by which we understand illness and health and is key to notions of vulnerability. The idea of the body as potentially open and vulnerable is not new: we see this idea in various cultural and temporal spaces.[20] When a particular society feels most vulnerable it is often the case that the permeability of individual bodies is experienced more acutely. Boddy's (1989) work reveals how body orifices (predominantly women's) are seen as vulnerable to outside forces when a community is under threat.[21] Perhaps the most influential scholar in this area is Mary Douglas, who has argued that bodily margins are analogous to social margins, and orifices are the "specially vulnerable points." Orifices, of course, are liminal in that they let what is in out and, more importantly for the purposes of this chapter, they allow what is out in. They are the points where substances in the environment can enter the body. In my book I discuss the shifting boundaries in military culture (that is, increased entrance of women and civilians, changing roles from warrior to peacekeeper), and suggest that the soldiers' notions of permeable barriers and vulnerable body boundaries are a reflection of their specific experiences.[22]

CHEMICALS: EVER-PRESENT, ONGOING RISK

The atmosphere is intimate and personal, but what happens when it is seen as entirely dangerous and risky? The Gulf War was seen as uniquely toxic—the environment itself was seen as harmful. The people I interviewed commonly referred to it as a "toxic soup." Not only were there individual toxins but there was the notion that these could be interacting in new and dangerous ways and, perhaps, creating even more terrifying substances. A particular experience can influence the way a person and/or group reacts to and engages with their environment in an ongoing way. Veterans feel themselves to be more sensitive to the atmosphere as a result of their Gulf War exposures. Sensitivity is linked with a distinction between natural substances and anthropogenic toxic substances—veterans suggest that their bodies react to anything unnatural. William, an ill veteran, was concerned about chemicals and exposures both in the war and in his daily life:

> I have now heard that as well as the inoculations, some of these things had additives put in. Now, I don't know what they were, but I do know that any chemicals in this day and age can damage the body. I believe that many chemicals in everyday use can damage the body. . . . I'm now using solvents at work; I always try to put a mask on and wear gloves so I don't come in contact with chemicals. I try to avoid foods that I know have a lot of additives. I always wash vegetables and fruits before eating them because I believe they spray them with chemicals, which I don't want to take. . . . I do believe chemicals do build up. I always have windows open. I feel there are a lot of chemicals and modern substances giving off vapors the whole time that are very bad and I think they will be proved in the future to be harmful to people's health.

One sufferer spoke about his sensitivity to chemicals: "I'm more sensitive now to cleaning stuff like bleach. Petrol fumes, as well, bring me back to the Basra Road. Especially burnt [smells remind me of] petrol bombs. Like when kids get a car and burn it out. It never affected me before, but it affects me now . . . more sensitive to it now." This veteran's comments are typical and demonstrate a belief that the Gulf War made them susceptible and vulnerable to everyday chemicals. Such discussion moves easily from this form of sensitivity to a different way one can be affected by the environment: in the form of smell. Smells are often spoken about as powerful reminders of the war. Indeed, discourse around post-traumatic stress disorder commonly refers to the way smells can be a trigger for flashbacks. Similar to a larger cultural trend of living life through a prism of fear, the fear of chemicals and the environment is one of the "quiet fears" of everyday life discussed by Furedi.[23]

Some veterans complain of multiple chemical sensitivity (MCS), and I have argued elsewhere that there is an overlap with MCS and other members of this family of emergent and contested conditions.[24] GWS is reflective of wider cultural health anxieties found at the time of its emergence. Recent times have been marked by a "resonance for an apparently endless series of health scares."[25] These anxieties have included radiation from mobile phones, MCS, and genetically modified organisms. The genealogy of these concerns themselves can be traced to cultural movements: anxieties about pesticides, for example, can be seen as linked to the ecology movement, heralded by Rachel Carson's seminal book *Silent Spring* (1962).

My anthropological project of seeking to understand GWS through contextualizing and tracing its roots resonates with the historical project of the scholars represented in this volume. Jim Fleming's body of work on historical readings of anxiety about climate change is of particular interest. Since the mid-1980s, the dominant concern about the environment has been global warming from rising concentrations of CO_2 and other greenhouse gases. As Fleming states, "In 1988 scientists James Hansen of the National Aeronautics and Space Administration announced to Congress and the world, 'Global warming has begun.' . . . [C]ombined with a continual stream of negative news about the stratospheric ozone layer since 1985, [this] has resulted in a major shift in humanity's relationship with the Earth's atmosphere. The clear blue sky now seems menacing."[26] As Jerome Namias pointed out in 1989, "the greenhouse effect is now firmly part of our collective angst."[27] Importantly, Hansen's comment was three years before the Gulf War, and such anxieties about the atmosphere and the environment more generally provide a backdrop to and inform their more specific concerns.

As Fleming has illustrated, the study of the "history of global change reminds us that there have been many global changes in our relationship to nature and that history, climate, and culture are closely interwoven. Climate apprehensions did not begin in 1988 or in 1957, or even in 1896."[28] What seems key here is that apprehensions about the environment have been found throughout history and in different cultural contexts and yet we seem to find ourselves in an acutely anxious time, which centers on the impact of humans on the environment. Fleming suggests that apprehensions "have been multiplying rapidly that we are approaching a crisis in our relationship with nature."[29] These dangers loop back, affecting our own health: feeding into illness beliefs and experiences.

More and more social problems are being examined through the prism of risk. Or, as Bill Durodié would say, we are less a risk society than a risk-perception society.[30] The concept of risk has become fundamental to the way both lay actors and technical specialists organize the world. In the contemporary Euro-American context, risk is impossible to ignore and is central to understanding health beliefs and behaviors. Three social scientists have written extensively on risk: the sociologists Ulrich Beck and Anthony Giddens and the anthropologist Mary Douglas. Beck and Giddens focus on the trend toward individualization in late modernity. Divorced from social solidarities, people feel vulnerable and see events as out of their control or inevitable.[31] For Beck and Giddens, risk has become reflexive or, in other words, humanity now has to deal with the new 'manufactured risks' of its own creation. Douglas's work focuses more on the social nature of decision-making in respect to risk. I would agree that the process by which people assess dangers is a social process, negotiated between individuals and institutions. Douglas's work on risk is a continuation of her work on purity, risk, and danger. She argues that distinguishing something as a risk is a way of making sense of the world as well as a method of keeping things in their proper place. Risk in our culture plays an equivalent role to sin or taboo, but it acts in the opposite way: it protects the individual against the community. Of course, it is not that chemical weapons are not a valid risk, but why are some risks singled out for particular dread?

It has been well established that popular perceptions of risks from environmental hazards are very different from scientific calculations of the same risk.[32] Why? Because beliefs and understandings about risk are cultural. The Lele people of former Zaire, for example, are exposed to a number of diseases and dangerous natural phenomena, but they emphasize lightning strikes, infertility, and bronchitis in their thinking about risk.[33] The expat community in which I lived in Qatar seems to emphasize anxiety about air conditioning and dust and yet quickly comes to terms with the perils of road accidents, although the latter are much more likely to cause immediate harm and death. This is particularly surprising given the statistics (which one can feel acutely on the streets of Doha), that 50 percent of young male Qatari deaths are due to road traffic accidents and that 18 percent of all deaths in Qatar in 2008 were caused by road accidents (compared to 2 percent in the United States).[34]

Interestingly, we again see air, in this case air conditioners, as the source of health anxieties. For British expatriate families the constant use and dependence upon air conditioners is a novel experience, which may in part account for the concern around their use. Similarly, the ever-presence of dust in the desert environment is unusual to those used to living in the United Kingdom and becomes threatening. However, once again we see that the nebulous and uncontrollable aspect of the air around us generates concern. Other groups and peoples emphasize other dangers. Not only are people's conceptions of risk often systematically distorted but at times they are also directly inconsistent, as when one and the same technology, namely radiation technology, is seen as being low-risk in medical use and high-risk in industrial use.[35] The concept of pollution is especially useful in political debates, because it bears the idea of moral defect.[36]

GWS appeared in a moment of heightened personal insecurities and risk aversion.[37] Veterans view the world as full of risk. They are likely to project pathological explanations for their symptoms. The popular belief is that the physical world is a potentially hostile place, full of chemicals, toxins, and viruses that erode health and well-being. In this way they are of their time, for "ours is a risk culture."[38] Importantly, it is the nature of the risk and the moral component that is key. For GWS the hazards at the center of their theories of causation are those that came from their own community.

Friendly Fire

The discourse surrounding the role of chemicals in the environment is political: who was responsible for the exposures in the war and who were those affected? In the case of GWS in the United Kingdom, the politics of responsibility is manifested in the elaboration of the friendly fire metaphor in which one's own body and one's own side are responsible for one's suffering. The issue of friendly fire was particularly acute in the Gulf War. While the death toll among Coalition forces engaging Iraqi combatants was very low, a substantial number of deaths were caused by accidental attacks from other allied units. Of the 148 American troops who died in battle, 24 percent were killed by friendly fire, a total of thirty-five service personnel. A further eleven died in detonations of allied munitions. Nine British service personnel were killed and eleven injured in a friendly fire incident when US Air Force Thunderbolt II A-10s mistakenly attacked two Warrior armored vehicles. The friendly-fire incidents held a prominent place in the imaginations of my informants and many spoke about them. The Gulf conflict, or more accurately, its aftermath, has extended the concept of friendly fire further. Some of the alleged "toxic" hazards that have been blamed for ill health are extensions of the friendly fire concept, since they also originated from "our side."[39]

Only one exposure—the oil fires—was explicitly the result of enemy action, and it is interesting to note that this is the one that has attracted the least coverage and controversy. The veterans I spoke to rarely mentioned this exposure as a concern. The controversy about the possible role of sarin seems the exception to the focus on allied responsibility in that the Iraqis had it, the allies did not; but any exposure from the Khamisayah incident is blamed on Coalition forces, and if we look further we see attempts to sue US contractors who are alleged to have supplied the Iraqi regime with the precursors needed to create sarin and cyclosarin.[40]

Of course, the concept of friendly fire, felt so acutely in the case of GWS, resonates with the fears discussed in the other chapters in this volume. Pollution, ozone depletion, airborne disease, and toxic *pneuma*: it is the atmosphere itself, ever-present and all encompassing, that becomes pathological. We need air to live and yet this life-saving substance can also be a risk to our health. I am reminded of Mary Douglas, who asks, "What are Americans afraid of?" The air they breathe. It is the invisibility of the risks that created anxiety among veterans. The negative effects of chemicals in the air and their effects on health are one way of making the invisible visible. But how does one fight that which one cannot see? We cannot. So instead, at least in the GWS case, an entity or a substance is focused upon as the enemy: the establishment or the chemicals. The veterans' fight for recognition of their disease became focused on the military, the government, and science. These three establishments are seen as responsible for making them ill and exposing them to the chemicals and toxins of the Gulf War. This concept of "friendly risk," where the agent is a product of their own side, resonates with ideas contained in other chapters in this volume, that the agent of pollution or climate change is perceived as a good thing, such as modernization, while faith in the values that unleashed these agents is undermined and a sense of betrayal created.

I am interested in the reasons why health scares often involve such impassioned debates: they represent a battleground upon which people could defend their views of the world and of their bodies. As the sociologist Émile Durkheim noted: who and what we fear, and how we express and act upon such fear, is constitutive of who we are. Anxiety about the effects of certain substances on health can be seen as part of taboo—a modern expression of these more basic ways of structuring our relation with the environment, and consequently this relation contributes to define our own identity. The discourse around causation of GWS certainly has the moral dimension of risk that Douglas emphasizes. Whereas chemical weapons play a role in theories of causation, they wane and instead exposures blamed on one's own side take center stage; when sufferers refer to their illness as "like friendly fire" they are emphasizing who is to blame. This is particularly interesting given the capacity of chemical weapons to terrorize and that the experiences of illness are significantly associated with a belief that chemical weapons were used. But when veterans talk about their illness chemical weapons diminish in importance. Instead, the preventative measures used to protect bodies from a chemical weapons threat are implicated foremost. The friendly fire metaphor is extended: one's body turns on itself in the illness just as one's own government has turned on one. Chemicals in everyday environments and bodies are the source of ongoing concern. Importantly, chemical weapons exposure would also only have affected a small number of soldiers and would not have accounted for the vast numbers of those ill with GWS (including those who were not deployed to the theater of war).

As Brenda Gardenour Walter's essay in this volume conveys, there is continuity between medieval fears and health anxieties of today where air and the environment are seen to affect and act upon the body. In the Middle Ages as well as now the atmosphere of a given region could become toxic enough to cause bodily harm. Why are these fears of a toxic environment so compelling and worthy of such dread, even among soldiers during warfare? These apparently new risks are not seen as part of the traditional military contract.[41] They fall outside what is expected and, of course, this is far more problematic when the hazards appear to be self-inflicted, as outlined in the discussion of friendly fire above and its particularly emotive resonance for the Gulf War. Control is also one reason why these risks are seen as particularly problematic. Like civilians, "the military seem accepting of other risks over which they feel they have a choice—such as driving or sports injuries, a perennial cause of serious injury and staffing difficulties."[42] The risks discussed here have the sense that they are uncontrollable. Furthermore, the invisibility of the risks and the resulting invisibility of the illness are central to this analysis.

Discussion of the Gulf War by GWS sufferers focuses on the atmosphere as dangerous and filled with toxins. There is an anxiety about chemicals and the substances to which they were exposed. Furthermore, these exposures change them irrevocably, damaging their bodies and altering their immunity, making them more vulnerable to potential hazards in their home environment. Their relationship with their bodies and their interaction with the environment around them is irrevocably altered by their experiences in the war. Their environment is risky and threatening.

NOTES

1. Tucker, *The Encyclopedia of Middle East Wars*, 470; "The Operation Desert Shield/Desert Storm Timeline," http://www.defense.gov/news/newsarticle.aspx-?id=45404.

2. See Kilshaw, "Friendly Fire," and Kilshaw, *Impotent Warriors*.

3. Kilshaw, *Impotent Warriors*.

4. Carden-Coyne, *Reconstructing the Body*; Jones et al., "Enduring Beliefs."

5. Wessely and Freedman, "Reflections on Gulf War Illness," 722.

6. Brewer et al., "Why People Believe."

7. "Studies Track Gulf War Illness," accessed June 2, 2012, http://www.gulflink.org/stories/disasternews/studies.htm.

8. Wessely, "Risk, Psychiatry."

9. Kilshaw, *Impotent Warriors*, 31.

10. Riddle et al., "Chemical Warfare."

11. Gray et al., "Postwar Hospitalization"; McCauley et al., "Illness Experience." Committee on Gulf War & Health 2004.

12. Riddle et al., "Chemical Warfare."

13. Kilshaw, "Paternity Poisoned."

14. The project is funded by the Qatar National Research Fund (award number: NPRP 4–1204–5-177).

15. Jones et al., "Enduring Beliefs."

16. Ibid.

17. Ibid.

18. Ibid.

19. Kilshaw, *Impotent Warriors*.

20. See Walter, this volume.

21. Boddy, *Wombs and Alien Spirits*.

22. Kilshaw, *Impotent Warriors*.

23. F. Furedi, "The Only Thing We Have to Fear Is the 'Culture of Fear' Itself," last modified 2007, accessed June 1, 2012, http://www.spiked-online.com/index.php?/site/article/3053/.

24. Kilshaw, *Impotent Warriors*.

25. Fitzpatrick, *MMR and Autism*.

26. Fleming, *Historical Perspectives*, 235.

27. Ibid., 238.

28. Ibid., 236.

29. Ibid., 3.

30. Durodié, "Concept of Risk."

31. Ibid.

32. Slovic et al., "Intuitive Toxicology."

33. Douglas and Wildavsky, *Risk and Culture*.

34. Qatar Statistics Authority, accessed June 2, 2012, http://www.qsa.gov.qa/eng/index.htm.

35. Slovic, "Perceptions of Risk."

36. Douglas and Wildavsky, *Risk and Culture*.

37. Furedi, "Only Thing We Have to Fear."

38. Giddens, *Modernity and Self Identity*, 3.

39. See Kilshaw, "Friendly Fire," and Kilshaw, *Impotent Warriors*.

40. Wessely and Freedman, "Reflections on Gulf War Illness."

41. Wessely, "Risk, Psychiatry."

42. Ibid., 464.

REFERENCES

Beck, Ulrich. *Risk Society: Towards a New Modernity*. London: SAGE, 1992.

Brewer, N. T., S. E. Lillie, and W. K. Hallman. "Why People Believe They Were Exposed to Biological or Chemical Warfare: A Survey of Gulf War Veterans." *Risk Analysis* 26 (2006): 337–45.

Boddy, Janice. *Wombs and Alien Spirits: Women, Men and the Zar Cult in Northern Sudan*. Madison: University of Wisconsin Press, 1989.

Carden-Coyne, Ana. *Reconstructing the Body: Classicism, Modernism, and the First World War*. Oxford: Oxford University Press, 2009.

Cohen, Stanley. *Folk Devils and Moral Panics*. New York: Routledge, 2002.

Douglas, M., and A. Wildavsky. *Risk and Culture: An Essay on the Selection of Tech-*

nological and Environmental Dangers. Berkeley: University of California Press, 1983.

Durodié, B. "Risk and the Social Construction of 'Gulf War Syndrome.'" *Philosophical Transactions of the Royal Society of London. Series B, Biological Sciences* 361, no. 1468 (2006): 689–95.

——. "The Concept of Risk." London: Nuffield Trust Discussion Paper, 2005.

Fitzpatrick, Michael. *MMR and Autism: What Parents Need to Know.* London: Routledge, 2004.

Fleming, James Rodger. *Historical Perspectives on Climate Change.* New York: Oxford University Press, 1998.

Giddens, A. *Modernity and Self Identity.* Cambridge: Polity Press, 1991.

Gray, G., T. C. Smith, J. D. Knike, and J. M. Heller. 1999. "The Postwar Hospitalization Experience of Gulf War Veterans Possibly Exposed to Chemical Munitions Destruction at Khamisayah, Iraq." *American Journal of Epidemiology* 150 (1999): 532–40.

Jones, E., I. Palmer, and S. Wessely. "Enduring Beliefs about Effects of Gassing in War: Qualitative Study." *BMJ* 335 (2007): 1313–15.

Kilshaw, S. "Friendly Fire: The Construction of Gulf War Syndrome Narratives." *Anthropology and Medicine* 11, no. 2 (2004): 149–60.

——. *Impotent Warriors: Gulf War Syndrome, Vulnerability and Masculinity.* Oxford: Berghahn Books, 2009.

——. "Paternity Poisoned: The Impact of Gulf War Syndrome on Fatherhood." In *Globalized Fatherhood: Emergent Forms and Possibilities in the New Millennium,* edited by M. Inhorn and W. Chavkin. Oxford: Berghahn Books, forthcoming.

Lupton, Deborah. "Introduction." In *Risk and Sociocultural Theory: New Directions and Perspectives,* edited by Deborah Lupton, 1–11. Cambridge: Cambridge University Press, 1999.

Martin, E. *Flexible Bodies: Tracking Immunity in American Culture from Days of Polio to the Age of AIDS.* Boston: Beacon Press, 1994.

McCauley, L., M. Lasarev, D. Sticker, D. G. Rischitelli, and P. S. Spencer. "Illness Experience of Gulf War Veterans Possibly Exposed to Chemical Warfare Agents." *American Journal of Preventative Medicine* 23 (2002): 200–206.

Riddle, J. R., M. Brown, T. Smith, E. C. Richie, K. A. Brix, and J. Romano. "Chemical Warfare and the Gulf War: A Review of the Impact on Gulf Veterans' Health." *Military Medicine* 168, no. 8 (2003): 606–13.

Slovic, P. "Perceptions of Risk: Reflections on the Psychometric Paradigm." In *Social Theories of Risk,* edited by S. Krimsky and D. Golding, 117–52. Westport, CT: Praeger, 1992.

Slovic, P., T. Malmfors, D. Krewski, C. K. Mertz, N. Neil, and S. Bartlett. "Intuitive Toxicology. II Expert and Lay Judgments of Chemical Risks in Canada." *Risk Analysis* 15 (1995): 661–75.

Tucker, S. *The Encyclopedia of Middle East Wars: The United States in the Persian Gulf, Afghanistan, and Iraq Conflicts.* ABC-CLIO, 2010.

Wessely, S., "Risk, Psychiatry and the Military." *British Journal of Psychiatry* 186 (2005): 459–66.

Wessely, S., and L. Freedman. "Reflections on Gulf War Illness." *Philosophical Transactions of the Royal Society of London Series B: Biological Sciences* 361, no. 1468 (2006): 721–30.

5

DECIPHERING THE CHEMISTRY OF LOS ANGELES SMOG, 1945–1995

Peter Brimblecombe

THE WAY CHEMISTS understand urban air pollution and how they relate this understanding to policymakers underwent great transformations in the twentieth century. These changes cannot be simplified to notions such as: "pollution got worse," because in many cities the concentrations of aggressive primary pollutants such as fly ash, smoke, and sulfur dioxide actually declined very substantially. The last, for example, decreased by more than an order of magnitude in some cities during the last half of the century, largely because coal was no longer burned in large quantities.[1] Changes in fuel sources—for example, from coal to petroleum—refocused pollution concerns from primary pollutants, like ash or smoke, to a more complex problem, that of the formation of photochemical smog. Nor can we glibly say that "chemists got smarter," for even as increasingly sophisticated scientific accounts emerged, it was the complex, nonlinear chemistry that failed to translate easily into policy recommendations for air quality or emissions standards.

Understanding Los Angeles smog was critical to forging the modern discipline of atmospheric chemistry. The way a variety of chemists and atmospheric scientists produced accounts of complex chemical reactions proved to be a mode of scientific development. Understandings of the LA smog have had far-reaching effects, well beyond petroleum-related pollution and beyond scientific inputs to policy debates. Without a history of the science behind an increasingly complex model of smog formation improvements in the air over LA and around, these developments cannot be properly understood.

Origins of Los Angeles Smog

The word "smog"—derived from the words "smoke" plus "fog"—was first coined by physician Henry Antoine des Voeux in 1905 in a paper that described pollution in British coal-burning cities. The term "smog" was used to describe several episodes in the early twentieth century where a smoke-like substance in the air not only compromised visibility but also caused a variety of adverse health effects. The most notable of these pre–World War II smogs was the St. Louis smog of 1939, where an inversion layer trapped thick smoke for over a week; streetlights were left on during the day to improve visibility. A few years later, during WWII, there were several episodes of severe air pollution in the city of Los Angeles; local residents referred to these as smogs. At the time some feared the pollution was the result of Japanese gas attacks; others thought it important not to reduce the output of strategic industry by imposing smoke control measures. Air pollution episodes were occasionally so bad in Los Angeles in the early 1940s that spectators in the stands could not see baseball games. However, the war effort did not obscure all local concerns; there were also worries about the health impacts of smog on Los Angeles residents in the summer of 1943.[2] Even in the 1940s there was an awareness of its smog's "peculiar nature" and the differences between it and smoke. Ironically, the term "smog" may have obscured the true picture, but this subtlety was lost on city authorities who set up a Bureau of Smoke Control in the health department in 1945.

After the war the *Los Angeles Times* brought air pollution expert Raymond R. Tucker from St. Louis to examine the smog problem and make recommendations. Tucker was a mechanical engineering professor at Washington University and the City Smoke Commissioner from 1934–37. He was credited with an accurate analysis and solution to the smog episode in St. Louis in 1939, and he would later rise to become mayor of the city. Tucker's *Los Angeles Times* report in 1946 about smog reinforced the notion that it was not just a few smokestacks causing the problem but a plethora of uncontrolled sources, including backyard rubbish incinerators and smoking trucks. However, Tucker's St. Louis experience was misleading in his investigation of LA smog. The air pollution in St. Louis derived from coal smoke, so it had few similarities with the photo-oxidative smogs of Los Angeles. Tucker's report recommended establishing unified air pollution control districts at the county level. On April 15, 1947, despite stiff opposition from industry, these recommendations became a reality.

Tucker's report did not single out the automobile as a source of LA's smog. He noted the presence of sulfur dioxide in the smog, but argued that the lack of sulfur in automobile fuel ruled the car out as a significant source of the smog. Researchers like Tucker had yet to recognize that LA smog was derived from evaporating components of fuels rather than industrial plants. Some early hints came from measurements of irritant aldehydes in road tunnels, but industrial sources received most of the attention. Since 1943 the problem had often been associated with the Aliso Street butadiene plant belonging to the Southern California Gas Company; the plant was temporarily shut down by public pressure in 1943. When the smog persisted, the plant, which produced important war materiel, was reopened. However, there was little justification in linking the smog to the butadiene plant at the time. [3] The Aliso Street Plant was used as a scapegoat, since both the media and county administrators wanted to identify a single villain that would allow easy remediation. Nevertheless, even Tucker's inaccurate *Los Angeles Times* report in 1946 discussed multiple sources. In 1947, I. A. Deutch, an assistant administrator at the Los Angeles County Office of Air Pollution recognized this facile attribution to a single source as an oversimplification.[4]

Determining the Photochemistry of Los Angeles's Air

By the 1950s it was more apparent that the pollution in Los Angeles was different from the earlier London or St. Louis smogs. Arie Jan Haagen-Smit, a Dutch biochemist at the California Institute of Technology studying air pollution's effects on plants, theorized that Los Angeles smog was caused by chemical reactions of automobile exhaust vapors in sunlight. He observed that the atmosphere of Los Angeles was rich in oxidizing substances and ozone. These could only come from reactions in the atmosphere, as they had no obvious primary sources. He developed these ideas in a series of papers that described the effects and were clear on the importance of the photochemistry of nitrogen oxides and the relevance of ozone.[5] He always stressed that there was "no resemblance at all to the problems of cities in the eastern parts of the United States," largely due to the unique geography of the LA basin and the number of automobiles there.[6]

Chemical understandings of smog remained limited even though it was easy to detect ozone in the air. Ozone suggested the existence of pathways for likely chemical reactions. However, scientists were unable to provide accounts of the reactions in any detail because they involved radicals, highly reactive molecules with unpaired electrons, which were unfamiliar in the 1950s. Still, Haagen-Smit saw the uniqueness and complexity of LA's smog and pushed for further research, which could lead to policy recommendations. He noted "a proper evaluation of the contribution of air pollutants to the smog nuisance must include not only the time and place of their emissions, but also their fate in air."[7] As early as the 1950s, Haagen-Smit's writings showed an early recognition of the subtlety required for control.

Public awareness of the novelty of LA smog grew in parallel with scientific understanding. A pictorial article in *Life* magazine in 1951 described the situation: "Now industrialization has caught up with Los Angeles to the extent that it has its own special brand of smog—less grim, but more eye-burning chemicals."[8] Smog also played a role in setting the stage in popular novels of the time, such as when the hard-boiled detective was pacing LA's fictional streets.[9] In 1953 Raymond Chandler wrote *The Long Goodbye* and shrouded his fictional detective Philip Marlowe in smog to set a suspenseful mood. Chandler later felt that others copied him and said "everybody is pissing about in the smog." Haagen-Smit also brought his account of smog to the public in *Scientific American* in 1964. He emphasized that "the Los Angeles atmosphere differs radically from that of most other heavily polluted communities."[10] However, as the use of the automobile spread and coal smoke became less dominant in urban atmospheres, photochemical smog became common in many more densely populated locations.

In the *Scientific American* article, Haagen-Smit argued for the need to reduce emissions from automobiles, and that "clean air can also be reached by dealing with both hydrocarbons and oxides of nitrogen." He suggested that reduction of each compound by half would achieve the required air quality, but this linearity belies the complexity in the system. Haagen-Smit was aware that various hydrocarbon components of the fuel would generate different smog formation processes (for example, olefins would create more smog than paraffins). As a result, changes in fuel composition could reduce smog production.[11]

Deeper scientific understanding of smog came from oxidative reaction sequences introduced in the late 1950s.[12] Chemists posited links between a variety of chemical pathways and processes, but could not fully clarify the details of the range of oxidation products. They had determined that aldehydes and peroxides were present as oxidation products, but had not yet established their effects on living tissue.[13]

Philip Leighton brought together several new developments in photochemistry in 1961 with the publication of his classic *Photochemistry of Air Pollution*. He was a professor of chemistry at Stanford University who had received many research grants to investigate the air pollution problem in California. Leighton's book on air pollution photochemistry was based on an enormous amount of new experimental evidence that his group had gathered in the preceding years. It includes an early mention of the hydroxyl (OH) radical, but without the clear understanding of its role that was to emerge by the 1970s. In his review of the book the theoretical chemist Henry Eyring reproduced Leighton's words: "a major share of the photochemically originated organic particulates in photochemical smog are due to the nitrogen dioxide-olefin photolysis and the reactions which follow." Eyring's review emphasized that it was not the initial nitrogen dioxide photolysis that was important; it was the subsequent reaction sequence that generated so many potentially reactive and harmful chemicals in the smog. Nevertheless the reactions proposed by Leighton may not have entirely convinced Eyring, who later wrote: "many features will undoubtedly be modified and amplified with time . . ."[14]

Morris Katz, a chemist for the National Research Council of Canada who pioneered air pollutant sampling and measurement methodology, also reviewed Leighton's book.[15] He realized the regulatory significance of the chemistry and wrote: "in order to control such harmful byproducts [oxidants, eye irritants and phytotoxicants] effectively, it is necessary to know the facts concerning their formation and reactions."[16] Katz also wrote a chapter of the World Health Organization's handbook on pollution published in 1961, in which he formulated new roles for the OH and HO_2 radicals.

In 1971 A. P. Altshuller and J. J. Bufalini dated the transition to a modern understanding of smog to the last half of the 1960s.[17] Still, Arthur C. Stern's major eight-volume work *Air Pollution*, the first three volumes of which were published in 1968, introduced a more advanced understanding than Leighton, but did not account for the key role of the hydroxyl radical.[18]

A more complete reaction sequence informed the report of atmospheric scientists K. Westberg and N. Cohen in 1969 and the work of Hiram Levy II in 1971.[19] In these writings Levy posited the OH radical as a basic ingredient for the production of photochemical smog.[20] Additionally, Westberg et al. showed the importance of reactions of OH with CO because of the high concentrations of CO in urban environments.[21]

Between 1945 and 1975, chemists and atmospheric scientists established the underlying reaction sequences that explained the relationship between photochemical smog, chemical processes involving volatile organic compounds, and sunlight. During these three decades scientists laid the basis of modern atmospheric chemistry and revealed the pivotal role of the OH radical in creating ozone and other oxidized compounds in the atmosphere. However, this knowledge came with a growing realization that the complexity imposed by these processes makes the creation of regulations to improve air quality much more difficult.[22]

THE TURN TO MODELING COMPLEX SMOG CHEMISTRY

Chemists and policymakers alike were enthusiastic about modeling the new photochemistry of smog and thought that models could be used to make rational policy decisions. However, developments in policy, while dependent on scientific understanding, do not necessarily move in lockstep with scientific developments. This was especially true in the case of photochemical smog once the complexities of smog formation were recognized. A number of chemists developed a range of approaches aimed at predicting changes in air pollution. Key figures in air pollution science, like John Seinfeld, typically brought together photochemical and atmospheric transport models. However, these approaches also required considerable simplification of both chemical and transport models, otherwise they were too large to code efficiently and too slow to run on existing computers.

In the 1970s, scientists began to recognize the importance of gasoline vapor (that is, volatile organic compounds or VOCs) as precursors to oxidants and as drivers in the production of ozone. Policymakers felt they needed descriptions of the actions of particular VOCs in order to develop regulatory policies and reduce smog. Because these compounds had not been isolated and analyzed, neither scientists nor policymakers understood their physical chemistry. Modeling required techniques that would allow chemists to estimate the rates of reactions that were not measurable. Some reaction rates could be measured experimentally in smog chambers, usually using gas chromatography or classical kinetic gas studies. Other rates could be approximated from thermochemical kinetics or molecular similarity.[23] Other methods involved lumping the chemistry of individual compounds together as imaginary idealized compounds that embraced whole groups of VOCs, yet many scientists remained unconvinced by such models.[24]

Pollution modelers often struggled with nonlinear interactions. The relationship of photochemical smog precursors to ozone or smog's effect on health concerns such as eye irritation is often nonlinear, making it hard to predict. The research of the 1960s had revealed that different compounds varied significantly in how rapidly they react in the atmosphere. Organic compounds originating from petrol vapor were involved in different reactions at different rates that could contribute to or inhibit ozone formation.[25] Pollution modelers tried to understand the overall outcome of reactions in this complex mixture of compounds in the atmosphere by defining reactivity in terms of the rates at which the organic compound reacts with the hydroxyl (OH) radical.[26] This made sense because most VOCs primarily react with OH radicals in the polluted atmosphere.

However, chemists W. P. L. Carter and R. Atkinson concluded, from collected kinetic data, that "reactivity scales based on OH radical rate constants neglected factors that affected the magnitude and the sign of the maximum incremental reactivity (MIR)." They argued "that it is a gross oversimplification to attempt to define a single reactivity scale for all organics that is applicable under all conditions."[27] R. G. Derwent and M. E. Jenkin went further and developed the MIR into the photochemical ozone creation potential (POCP), which took multiple fates and degradation pathways of VOCs into account.[28] The POCP also provided a way to compute the net ozone production in subsequent reactions. Derwent and Jenkin related everything to ozone-producing, highly reactive ethene (ethylene), which was given the maximum score of 100 on the POCP scale. Modeling tactics like the POCP helped overcome the limitations of the linear approach that Haagen-Smit had taken thirty years earlier.

History of Air Pollution Policy

Although there are examples of attempts to regulate air pollution in classical and medieval times, it was only in the nineteenth century that substantial policies emerged to address air pollution through emission control. Regulators sought to control the fuels that were burned, to provide specifications for furnaces and chimneys, and to shift the location of industrial activity. Emission control is the most obvious approach as it benefits from being able to address a direct link between the polluter and the emissions. It is also relatively simple because it doesn't involve a large amount of environmental monitoring, and where monitoring is required it is often limited to a few measurements of exhaust gases.

Improvements in emissions from individual sources, although promising, did not necessarily result in a great improvement in overall air quality, especially where the number of sources was rapidly increasing. In Europe and North America there were fatalities during air pollution episodes in the Meuse Valley (Belgium, 1930), Donora (Pennsylvania, 1948), and most significantly London (the Great Smog of 1952). Extreme events such as these tended to drive government investigations and subsequent regulation, so there was a tendency for administrators to focus on these rare events rather than the day-to-day impact of air pollutants.

Nevertheless, air pollution episodes such as these resulted in more effective urban monitoring networks, which until the 1950s tended to record the concentrations of one or two pollutants at just a few sites or even simply the total mass of solid material deposited from the atmosphere. Monitoring networks allowed regulation to shift from a focus on lowering emissions to lowering the ambient concentrations. This approach, which was reasonably successful, was aimed at reducing the concentrations of pollutants to levels that health experts believed would protect urban populations. The approach remains in evidence today, as we still express air quality standards in terms of allowable concentrations, although these are now typically associated with an exposure time. Despite the move to think of air quality in terms of concentrations, policymakers presumed that there was a linear relationship between pollutant emissions and concentrations and tried to reduce primary emissions accordingly. The complexities of photochemical smog meant that this presumption was incorrect, so "air quality management" was developed as a technique to control the new forms of secondary air pollution that have emerged since the 1940s.

CHEMISTRY AND CONTEMPORARY POLICY

Managing smog air pollution is clearly a much more difficult problem than the first smoke abatement enthusiasts in Victorian England might have imagined when they grappled with the coal smoke that filled the air of their cities. Victorian sanitary scientists recognized a clear association between the polluter—for example, a smoking chimney—and visible pollution in the air. In contrast, photochemical smog presented twentieth-century regulators and chemists with a problem of much greater complexity, where the link between the components of smog and the resultant pollution was mediated by atmospheric chemistry. Other air pollution phenomena have been like coal smoke in their clear relationships between pollutants and pollutions, such as the effect of CFCs/freons on stratospheric ozone depletion (and their replacement by HCFCs, which also have predictable environmental effects).[29]

The environmental Kuznets curve, a graphical representation of economist Simon Kuznets's (1901–85) theory of industrial and economic development, provides a useful heuristic tool describing simpler pollution relationships, such as the coal smoke situation.[30] The theory predicts that the growth of industrialization pollutants will be balanced by economic growth, providing societies with the resources to reduce pollution as the environmental costs exceed socially acceptable levels. There are no a priori reasons why the economy or environment should improve once per capita income reaches a certain level; the Kuznets curve simply predicts this development. The theory has been applied successfully to a variety of primary local pollutants. Some have argued that the changes over time shown in the Kuznets relationship may come not so much from economic issues but rather from the development of pollution-regulating institutions.[31]

The fuel–air pollution–regulatory nexus is extremely dynamic, and the types of fuels burned for urban transportation may be about to undergo another dramatic change. The contemporary interest in the use of biofuels may introduce a new range of volatile organic compounds into the atmosphere. The view has often been that these are less polluting than petroleum products.[32] However, there are fears that biofuels will yield higher nitrogen oxide emissions.[33] Furthermore, transitioning to the biofuel methanol has the potential to produce higher formaldehyde concentrations in urban atmospheres. Other changes are also underway that alter the regulatory response. There is a balance between global pollution issues, such as greenhouse gases and local pollutants, and the convenient assumption is that it is a win-win situation and that decreasing local emissions may also decrease greenhouse warming. While carbon dioxide contributes to the greenhouse effect and warms the atmosphere, sulfate particles produced from the oxidation of sulfur dioxide from coal burning may cause a cooling effect, known as global dimming. This may be overly simplistic, since it ignores the complex interactions of global pollutants.[34] Thus, reducing coal smoke yields important improvements in air quality, but the concomitant reduction in sulfate particles in the atmosphere might enhance global warming.

In addition to such complex chemical interactions, models must take into account the global transport and fate of pollutants. Many regions have unique air pollution issues based on their geographical settings. For example, there is a complex chemistry and global transport of dust, especially over the Asia Pacific Rim, where in the spring dust is blown from the central deserts of Mongolia and China. In coastal environments air pollutants mix with sea salts and allow for an elaborate chlorine chemistry. In volcanic regions of the world, such as Hawaii, "vog," or volcanic-fog, has become a health issue. Forests release VOCs, such as isoprenes and terpenes, which oxidize to form ground-level ozone. Massive forest fires over southeast Asia produced large plumes of smoke, especially noticeable in the strong El Niño of 1997. It seems likely that organic compounds absorb and react on desert dusts as they pass over urbanized coastal China, that forests produce complex and highly reactive biological volatile organic compounds, and that smoke from forest fires reacts chemically in complex, nonlinear ways. Assessment and modeling of these atmospheric phenomena depend on improved understanding of the chemistry of the atmosphere that first arose from research into the chemistry of Los Angeles smog.

The complex interactions between different chemical reactions in the atmosphere were relevant to political decision-making in the management of air pollution in the late twentieth century. In particular, these models made regulation more difficult because nonlinear models failed to single out the worst pollutants and polluters for sanctions.

Air pollution regulation has evolved into air quality management techniques that extend past the simple emission controls of the past.[35] Pollution modeling for policy development has fostered interest in a range of economic techniques for controlling air pollutants—for example, emissions trading.[36] Choosing effective management strategies is not always easy because of increasingly complex accounts of air pollution chemistry. Neither politicians nor the public seem to feel especially comfortable with chemistry. However, translating scientific knowledge into a public arena is hardly a new problem.

In the twentieth century, air pollution was no longer visibly linked with a single source as it was in the days of the smoky factory chimney. Chemists developed new understandings of air pollution that depended more on chemical transformations in the atmosphere than a direct coupling of pollution to its sources. These accounts have had profound effects on regulatory regimes. The pioneering studies of new atmospheric chemistry focused on smog formation in Los Angeles. After World War II, Arie Haagen-Smit, like others who worked on the emerging complexity of photochemical pollutants, realized that these new theories needed to be incorporated into policy. The key problem for regulators was that pollution found in the atmosphere had become detached from the direct emissions from automobiles and industry. The secondary photochemical pollutants in Los Angeles smog could no longer be attached to a single source as chemical reactions now mediated the resultant pollution. This disrupted the simple link between emission and pollutant concentration and required atmospheric science and modeling to become part of the regulatory process. This led to the development of techniques known as "air quality management" in the last decade of the twentieth century.

There was, almost from the very beginning, a commitment to communicating these ideas to both policymakers and the public. Thus although smog problems have taken many decades to improve, such delays and gradual changes reflect more the difficulties and complexity inherent in the problem than a lack of engagement by scientists with policy. Atmospheric scientists have been notably successful in tackling acid rain and the ozone hole.[37] However, anthropogenic climate change, which shares the feature of chemical and political complexity with LA's smog, remains a challenge for the twenty-first century.

NOTES

1. Brimblecombe, *Big Smoke*, chapter 6; Brimblecombe, "Clean Air Act," 311–14.

2. Dunsby, "Localizing Smog," 185; Krier and Ursin, *Pollution and Policy*, chapter 3.

3. Butadiene (C_4H_6), an important industrial hydrocarbon used in the production of synthetic rubber.

4. Dunsby, "Localizing Smog," 184.

5. Haagen-Smit, "Chemistry and Physiology"; Haagen-Smit, Bradley, and Fox, "Ozone Formation."

6. Haagen-Smit, "Air Conservation," 869–78.

7. Ibid.

8. "How Smog Hurts," 61–63.

9. Brimblecombe, "Writing on Smoke," 93–114.

10. Haagen-Smit, "Control of Air Pollution," 24–31. In fact, these types of smog are not unique to Los Angeles and are found in cities using large amounts of volatile liquid fuels and where sunlight is strong enough to promote photochemical reactions.

11. Haagen-Smit, "Urban Air Pollution," 1–18.

12. Haagen-Smit and Fox, "Ozone Formation," 1484–87; Johnston, "Photochemical Oxidation," 1488–91; Littman, Ford, and Endow, "Formation of Ozone," 1492–97; Mader et al., "Effect of Present Day Fuels," 1508–11.

13. Haagen-Smit, "Chemistry and Physiology," 1346.

14. Eyring, "Sunlight and Health Hazards."

15. Katz, "Photochemistry," 878.

16. Katz, "Some Aspects," 153.

17. Altshuller and Bufalini, "Photochemical Aspects," 39–64.

18. Stern, *Air Pollution*.

19. Westberg and Cohen, "Chemical Kinetics"; Levy, "Normal Atmosphere," 141–43.

20. Ibid.

21. Westberg, Cohen, and Wilson, "Carbon Monoxide," 1013–15.

22. See chapters by Dunn and Johnson, and by Lee, this volume.

23. Benson, *Thermochemical Kinetics*; Darnall, Atkinson, and Pitts, "Rate Constants," 1581–84.

24. Heymann, "Lumping," 318–32.

25. Altshuller and Bufalini, "Photochemical Aspects," 39–64.

26. Darnall et al., "Reactivity Scale," 692–96.

27. Carter and Atkinson, "Experimental Study," 670–79.

28. Derwent and Jenkin, "Hydrocarbons," 1661–78.

29. Bowden and Brimblecombe, "Haloacetic Acids," 285–91.

30. Halkos, "Environmental Kuznets Curve," 581–601.

31. Bhattarai and Hammig, "Institutions," 995–1010.

32. Texas State Energy Conservation Office, "Biodiesel Air Quality," http://www.seco.cpa.state.tx.us/re_biodiesel-air.htm.

33. Brooke Coleman, *Renewable Energy Action Project*, http://www.nebiofuels.org/pdfs/AQ_Summary.pdf.

34. Stanhill, "Perspective on Global Warming," 58.

35. Elsom, *Atmospheric Pollution*, chapter 7.

36. See Polli, this volume.

37. See Rothschild and Dörries, this volume.

REFERENCES

Altshuller, A. P., and J. J. Bufalini. "Photochemical Aspects of Air Pollution: A Review." *Environmental Science and Technology* 5 (1971): 39–64.

Benson, S. W. *Thermochemical Kinetics: Methods for the Estimation of Thermochemical Data and Rate Parameters.* New York: John Wiley & Sons, 1973.

Bhattarai, M., and M. Hammig. "Institutions and the Environmental Kuznets Curve for Deforestation: A Crosscountry Analysis for Latin America, Africa and Asia." *World Development* 29 (2001): 995–1010.

Bowden, D., and P. Brimblecombe. "Haloacetic Acids in the Environment—Is the Cure for CFCs Worse than the Disease?" *Progress in Environmental Science* 1 (1999): 285–91.

Brimblecombe, P. *The Big Smoke.* London: Methuen, 1987.

——. "Writing on Smoke." In *Dirty Words*, edited by H. Bradby, 93–114. London: Earthscan, 1990.

——. "The Clean Air Act after Fifty Years." *Weather* 61 (2006): 311–14.

——. "Transformations in Understanding the Health Impacts of Air Pollutants in the 20th Century." *European Physical Journal Special Topics* 1 (2009): 47–53.

——. "Air Pollution and Society." *European Physical Journal Conferences* 9 (2010): 227–32.

Brimblecombe, P., and Carlota M. Grosi. "Millennium-Long Damage to Building Materials in London." *Science of the Total Environment* 407 (2009): 1354–61.

Carter, W. P. L., and R. Atkinson. "An Experimental Study of Incremental Hydrocarbon Reactivity." *Environmental Science and Technology* 21 (1987): 670–79.

Darnall, K. R., R. Atkinson, and J. N. Pitts Jr. "Rate Constants for the Reaction of the OH Radical with Selected Alkanes at 300 K." *Journal of Physical Chemistry* 82 (1978): 1581–84.

Darnall, K. R., A. C. Lloyd, A. M. Winer, and J. N. Pitts Jr. "Reactivity Scale for Atmospheric Hydrocarbons Based on Reaction with Hydroxyl Radical." *Environmental Science and Technology* 10 (1976): 692–96.

Derwent, R. G., and M. E. Jenkin. "Hydrocarbons and the Long-range Transport of Ozone and PAN across Europe." *Atmospheric Environment* 25 (1991): 1661–78.

Dunsby, J. "Localizing Smog: Transgressions in the Therapeutic Landscape." In *Smoke and Mirrors: The Politics and Culture of Air Pollution*, edited by E. M. Depuis, 170–200. New York: New York University Press, 2004.

Elsom, D. M. *Atmospheric Pollution.* Oxford: Blackwell, 1992.

Eyring, H. "Sunlight and Health Hazards." *Science* 135 (1962): 427–28.

Haagen-Smit, A. J. "Chemistry and Physiology of Los Angeles Smog." *Industrial & Engineering Chemistry* 44 (1952): 1342–46.

——. "Air Conservation." *Science* 128 (1958): 869–78.

——. "Urban Air Pollution." *Advances in Geophysics* 6 (1959): 1–18.

——. "The Control of Air Pollution." *Scientific American* 210 (Jan. 1964): 24–31.

Haagen-Smit, A. J., C. E. Bradley, and M. M. Fox. "Ozone Formation and the Photochemical Oxidation of Organic Substances." *Industrial & Engineering Chemistry* 45 (1953): 2086–89.

Haagen-Smit, A. J., and M. M. Fox. "Ozone Formation in Photochemical Oxidation of Organic Substances." *Science* 48 (1956): 1484–87.

Halkos, G. E. "Environmental Kuznets Curve for Sulfur: Evidence Using GMM Estimation and Random Coefficient Panel Data Models." *Environment and Development Economics* 8 (2003): 581–601.

Heymann, M. "Lumping, Testing, Tuning: The Invention of an Artificial Chemistry in Atmospheric Transport Modeling." *Studies in History and Philosophy of*

Science Part B—Studies in History and Philosophy of Modern Physics 41 (2010): 318–32.

"How Smog Hurts Lungs." *LIFE*, Feb. 12, 1951, 61–63.

Johnston, H. S. "Photochemical Oxidation of Hydrocarbons." *Science* 48 (1956): 1488–91.

Katz, M., "Some Aspects of the Physical and Chemical Nature of Air Pollution." In *Air Pollution*, 97–158. Geneva: World Health Organization, 1961.

———. "Photochemistry of Air Pollution." *American Journal of Public Health* 52 (1962): 878.

Krier, J. E., and E. Ursin. *Pollution and Policy.* Berkeley: University of California Press, 1977.

Leighton, P. A. *Photochemistry of Air Pollution.* New York: Academic Press, 1961.

Levy, H., II. "Normal Atmosphere: Large Radical and Formaldehyde Concentrations Predicted." *Science* 173 (1971): 141–43.

Littman, F. E., H. W. Ford, and N. Endow. "Formation of Ozone in the Los Angeles Atmosphere." *Science* 48 (1956): 1492–97.

Mader, P. P., M. W. Heddon, M. G. Eye, and W. J. Hamming. "Effect of Present Day Fuels on Air Pollution." *Science* 48 (1956): 1508–11.

Schuck, E. A., and G. J. Doyle. *Photooxidation of Hydrocarbons in Mixtures Containing Oxides of Nitrogen and Sulfur Dioxide.* Report No. 29. San Marino, CA: Air Pollution Foundation, 1959.

Spedding, D. J. *Air Pollution.* Oxford Chemistry Series. Oxford: Oxford University Press, 1974.

Stanhill, G. "A Perspective on Global Warming, Dimming, and Brightening." *Eos* 88 (2007): 58.

Stern, Arthur C. *Air Pollution.* 3 vols. New York: Academic Press, 1968.

Westberg, K., and N. Cohen. *The Chemical Kinetics of Photochemical Smog as Analyzed by Computer.* Report ATR-70-(8107). El Segundo, CA: Aerospace Corporation, 1969.

Westberg, K., N. Cohen, and K. W. Wilson. "Carbon Monoxide: Its Role in Photochemical Smog Formation." *Science* 171 (1971): 1013–15.

6

CHASING MOLECULES

Chemistry and Technology for Automotive Emissions Control

Richard Chase Dunn and Ann Johnson

What air could be more obviously toxic than the tailpipe emission of an automobile? Everyone knows automobiles produce hazardous air pollution, whether as the components of visible smog or by emitting invisible, deadly carbon monoxide, or as other less well-known emissions. Yet many adults in the industrialized world depend on or even love their cars; pollution has not frightened most drivers out of their vehicles. The challenge of reducing air pollution without reducing miles driven is real, and engineers have been addressing it since the 1950s. However, the story of emissions reduction is not a simple one in which scientists have explained the chemical phenomena and engineers designed the devices that intervene in producing the targeted chemicals. As Peter Brimblecombe's chapter in this volume shows, scientific understandings of air pollution, particularly smog, have changed since the 1940s. Emissions control devices have coevolved with scientific explanations of pollution. This chapter examines the ways that changing understandings of air pollution have driven automobile engineers to address a long and changing list of chemical compounds emitted by internal combustion engines with a dynamic array of technological fixes with the goal of making a less polluting car. Engineers have had great success in making vehicle emissions less hazardous, but in the twenty-first century have we reached the limits of making the internal combustion engine cleaner?

Cars are real polluting machines; this is not a specious allegation. As early as 1966, the United States attributed 86 million tons of air pollutants to automobiles, out of a total of 146 million tons—over half of America's identified air pollutants.[1] Of the main categories of pollutants cars produce—carbon monoxide, hydrocarbons, and nitrogen oxides—automobiles produce the majority of all three, which have negligible nonanthropogenic sources. Automobile-related efforts to improve air quality since the late 1950s have focused primarily on devices to capture evaporating hydrocarbons, improve the combustion efficiency of the engine, and capture noxious tailpipe emissions. As a result, the three main pollutant classes have all been significantly reduced since 1965, when the first federal vehicle emissions standards were set. However, the vehicular emission of carbon dioxide—a seemingly harmless "natural" component of air—has increased due to the increased efficiency of the combustion process in cars. Whether carbon dioxide can be addressed as the other pollutants have been remains an open question and will most likely depend on reducing combustion as an energy source in the car, as hybrids and electrical vehicles do. Even with the carbon dioxide question open, it remains useful to understand how progress has been made in reducing the other noxious emissions of the internal combustion engine.

The development of technologies to reduce carbon monoxide, hydrocarbons, and nitrogen oxides has been dependent on three general factors, all of which involve research that is only indirectly related to the design of emissions control technologies. First, scientific understandings of the thermodynamic and chemical processes taking place in the car's engine and exhaust system have themselves been evolving since the 1960s; combustion cycles have become better understood. Secondly, atmospheric chemists began to investigate air pollution and understand what caused smog; this dynamic understanding fed into regulatory regimes which came to define what a pollutant was and what the penalties for its production would be. Thirdly, the automobile has never been a closed technology, and its development has depended on other technologies developed externally to the automobile, such as microprocessors, to manage and manipulate the production of a variety of combustion products and by-products. The interactivity of these three factors in engineering emissions control devices yields a complicated story of science, technology, and regulation. What's more, emissions control technologies that have been developed since the late 1950s are also interactive—they affect how each other works. This chapter shows many of these entanglements.

The entangled story of science, technology, and regulation told here is different from a linear account of science begetting technology in many ways. The simple story that all these emissions control technologies were predicated on a particular chemical understanding of the photochemical process of smog formation is incomplete to the point of being inaccurate. What might be called sound science (but might better be called complete science) was neither necessary nor sufficient for the development of intervening technologies. In fact, chemical understanding of smog's composition and formation developed alongside the technologies to prevent the production of smog's chemical components. Engineers and chemists worked in parallel, with better communication at some junctures than others. Since smog chemistry was an active and dynamic field, it was not unusual for automobile engineers to base their work on out-of-date chemistry. What is surprising is that these out-of-phase developments seem to have few negative consequences; reduction of the combustion by-products, except for CO_2 and water, proved to be critical and was accomplished regardless of the completeness of the account of photochemical smog production. Cars are clearly cleaner today, in the sense of emitting fewer smog-forming compounds, than they were in the 1960s. However, the carbon dioxide they produce is inescapable—combustion produces CO_2, and more efficient combustion optimizes production of CO_2.

In this chapter, we also show that engineers used a wide variety of technologies and tactics to track and usually transform the undesirable molecules in automobile emissions. Engineers were also important participants in the development of air quality standards and in regulatory regimes such as CHESS, the EPA's monitoring and research program (see Lee, this volume). Engineers and their professional societies were also involved in debates about the relative balance between government regulation and voluntary action by industry and consumers. By seeking technological fixes for air pollution, they were trying to mediate the connection between cars and smog, and thereby lessen the social burden involved in cleaning up the atmosphere. The development of successful technologies to capture or catalyze unwanted by-products of petrochemical combustion shows cycles of technological fixes, where technologies begat other technologies in an attempt to keep up, even belatedly, with evolving chemical understandings of atmospheric pollution.

The story of Arie Haagen-Smit's work at the California Institute of Technology to determine the chemical composition of Los Angeles's infamous smog is fairly well known.[2] In the years during and just after WWII, Haagen-Smit—a plant biochemist by trade—was trying to condense pineapple essence in order to analyze and synthesize it in his Pasadena lab. He passed the fumes through a large volume of ambient air and then into a trap with liquefied air in order to yield a condensed, frozen vapor of the pineapple essence. The vapor included some curious brown drops, which Haagen-Smit determined analytically to be a peroxy-organic substance. Haagen-Smit quickly discovered that the compound he had inadvertently isolated with the frozen pineapple vapor originated in petroleum combustion.[3]

A lengthy public-funded study had preceded Haagen-Smit's research to determine the composition of Los Angeles's growing and easily visible smog. The prevailing wisdom was that LA's pollutant was sulfur dioxide being aerosolized into sulfur trioxide, the chemical culprit in St. Louis's toxic fog episode in 1939.[4] The WWII-era LA smog study preceded the tragic London "Great Smog" episode of 1952, which did involve the same chemical compounds as St. Louis's. These incidents were caused by the emissions from burning low-grade "soft" coal and the sulfur compounds being trapped by atmospheric inversion layers over the cities. One prevailing theory prior to Haagen-Smit's work was that LA's "sulfurous smog" was (largely) a product of the butadiene plants created during WWII. Yet there were doubters, including chemical instrument maker Arnold Beckman.[5] At the time, Beckman was Haagen-Smit's colleague on the faculty of the California Institute of Technology. Haagen-Smit, intrigued by the noxious brown drops, took a leave of absence from Caltech to investigate the matter. His training as a chemist was important to his approach to the problem, and his wife, Zus, explained in an oral history, "He painstakingly delineated the photochemical process by which organic materials in the air—mostly hydrocarbons—are oxidized through the combined actions of oxides of nitrogen and sunlight to become smog."[6] Research on automobile emissions followed quickly, in part by an unsuccessful effort to disprove Haagen-Smit's theory. While oil and gas refineries sought ways to reduce the off-gassing of unburned hydrocarbons (because it indicated a loss of their product), automobile manufacturers fought accusations that their products were the causes of smog, since they worried from the start that there were no economic incentives for car buyers to spend more money to purchase cars equipped with new technologies to control emissions.[7] Even a buyer motivated to spend money for reduced emissions faces a situation in which the one car he or she drives contributes such a miniscule part of the emissions load that the increased price is not justified, if it is at all significant. Automobile engineers discovered in this research that the production of the ingredients of smog was complicated and entangled. Haagen-Smit's research indicated only what compounds were problematic—his research focused on the process in the atmosphere, not in the car's engine. Changing accounts of smog would target new molecules and would change the relative emphasis different compounds received as agents of blame.

Chemically, looking at the automobile, the problem may have appeared simple in that the three groups of chemical compounds involved were hydrocarbons from petroleum products, carbon monoxide and carbon dioxide produced in fuel combustion, and nitrogen oxides formed from the by-products of hydrocarbons burning in nitrogen-filled ambient air. However, the interactions of these gases in the engine and exhaust system were not simple, nor was their interaction isolatable from the performance of the engine. The thermodynamics of combustion were part and parcel of the chemical story. Making a car produce fewer smog-creating components could not be reduced to an easy, two-parameter optimization problem although engineers typically prefer such a reduction. When emissions controls failed to be framable in that manner, engineers had to work more empirically—trial and error. Facing the challenges of optimization is where the actual story diverges from the simple story that is often told of scientific understanding leading the way to technological fixes. Here the technological fixes turn out to be in conflict with each other and at odds with the users' expectation of reliable performance. The automobile emissions story would be short and linear if it were simply the case that knowing what compounds caused the problem would outline a clear path to solutions. In fact it is a long one, and involves nothing less than a complete transformation of the car into a fuzzy-logic-controlled, semi-conductor-dependent system of systems.

Technological Fixes Part I: Unburned Hydrocarbons

The one piece of the puzzle that Haagen-Smit's research laid out clearly was the problem of unburned hydrocarbons. This was the problem that oil refineries faced as well. The refineries could simply redesign their tanks to eliminate the air volume between the oil surface and the vessels' roofs with a floating cap.[8] For the automobile manufacturers, there also seemed to be a manageable modular retrofit: a device that pulled the engine's unburned vapors back in. Another cycle of burning promised to eliminate a large share of the emissions and would affect little else in the engine. They were even able to modify a device invented during the war to prevent tank engines from being flooded when tanks drove through deep water: the positive crankcase ventilation (PCV) valve, which circulated fresh air through the crankcase and carried blow-by gases into the intake manifold for reburning. In theory, this also increased the efficiency of the engine (though at levels probably undetectable by the user). Prior to 1960, cars had a road draft tube to remove any fuel-air mixture or vaporized fuel that escaped from the pistons into the crankcase. These vapors were called blow-by gases. The road draft tube fed the blow-by gases from the crankcase into the atmosphere by allowing air to flow past the open oil filler cap that led to a small vacuum that sucked the blow-by gases out of the crankcase and released them.[9] Doing so also increased the engine's life by allowing it to run cooler and at a lower pressure. Replacing the road draft tube with a PCV valve would eliminate unburned hydrocarbons by approximately 20 percent, but blow-by gases still needed to be removed from the crankcase. In 1961, the state of California passed legislation requiring PCV valves on cars beginning with the 1963 model year. Federal mandates would follow with the 1966 model year. PCV valves were credited with reducing hydrocarbons by 12 percent in just three years, between 1965 and 1968.[10] This would be the first—and probably the last—simple technological fix in the search for a cleaner-burning car. Discussion of these problems constitutes the earliest emissions-related communications at the annual meetings of the Society for Automotive Engineers (SAE), a large US-based professional society concerned with all parts of transportation. The tailpipe exhaust of an automobile also contains unburned hydrocarbons, but reducing those was more complicated than simply recirculating blow-by gases. Cold engines (engines that were just started) released higher levels of partially burned hydrocarbons than hot engines. Consequently, the fraction of hydrocarbons that escaped unburned depended on the temperature of the engine; as the engine warmed up, this fraction declined. Short trips with many stops (a common enough driving pattern) allowed the engine to cool and maximized unburned hydrocarbon output. In addition, the ratio of fuel to air also affected the amount of unburned hydrocarbons. A "rich" mixture of more fuel to less air increased hydrocarbon emissions, but it could also improve performance. A lean mixture lowered hydrocarbons but could also cause the engine to misfire, which would release a lot of unburned hydrocarbons. As a result it was well known that valve timing (which affects engine heat) and carburetion could be used to minimize hydrocarbon output, but this would put the onus on the car owner to keep a vehicle well tuned, and drivers weren't always able to determine when there was a problem. The real issue was that valve timing and carburetion could not be varied dynamically to optimize the engine's performance as its temperature varied. A solution to this problem, developed by engineers in the 1960s, was called the thermostatic air cleaner (TAC) system.[11] The TAC system allowed a lean mixture to yield good performance in a cold engine (<100°F). The designers' goal was to obtain the most complete combustion and limit the output of both unburned hydrocarbons and carbon monoxide, a by-product of the combustion of hydrocarbons in the cylinders. When the engine was cold, the TAC circulated air warmed in the exhaust manifold to the carburetor. When the engine warmed up (>130°F), the TAC stopped operating and normal air circulation would resume.

Air injection reaction (AIR) systems also focused on creating more complete combustion, but these systems focused on the exhaust manifold.[12] AIR systems allowed the hydrocarbons and carbon monoxide to continue to burn in the exhaust manifold, converting the harmful compounds into carbon dioxide and water. AIR systems began to appear on vehicles in 1966 and were common by 1968. The AIR system was an exhaust modification, so it worked independently from the combustion process modified by the TAC. Additionally in the late 1960s there were some new technologies added to the fuel tank and lines to reduce fuel evaporation.[13] This kind of modularity was highly prized by automotive engineers, but could not always be achieved in practice.

The passage of a series of laws in California in the 1960s precipitated similar legislation at the federal level in the late '60s and early '70s. California's legislation had encouraged innovation by forcing automotive manufacturers to modify their models to continue to serve the lucrative California market. As more and more devices were produced, economies of scale brought down the prices of these devices, which were simply stuck onto engines already in production. The federal Motor Vehicle Air Pollution and Control Act (1965) allowed the Department of Health, Education, and Welfare to set federal emissions standards, limiting the allowable hydrocarbon emissions of a new car in model year 1968 to 275 parts per million (ppm) in the exhaust stream. This set the stage for the better-known Clean Air Act of 1970 and the creation of the Environmental Protection Agency by Richard M. Nixon. The jurisdictional challenges California faced in regulating air quality sometimes translated to the federal level but often did not. Figuring out whose responsibility various pollutants were proved challenging. Figuring out how to enforce regulation would be a looming problem of the 1970s.

Because the TAC and AIR were, in many cases, proprietary and covered by several patents, there was little direct discussion of them at SAE meetings. Different automobile companies developed their own versions. Discussions of how to measure the efficacy of these devices were, however, commonplace at SAE meetings in the 1960s. Measuring the emissions of an automobile is a more difficult problem than it seems.[14] The amount and composition of automotive emissions varies with the relative temperature of the engine, the speed of the car, the type of driving (highway versus stop-and-go), the composition of the fuel, and even the altitude and composition of the ambient air. Initially, emissions measurements were made by placing the car on a dynamometer with a probe in the tailpipe or a suction system at the point of the escaping blow-by gases. Dynamometers and tailpipe probes are still the preferred instruments for annual emissions inspections in the states that require them. However, engineers realized that these measurements were not accurate and could be misleading in both over- and underestimating how a given vehicle was performing.[15] On-street monitoring was the only way to get accurate results for the changing emissions under real driving conditions, but that measured only the idiosyncratic output of a single car in a particular location and would vary widely. The seemingly standardized garage output of the tailpipe probe in a stationary vehicle was misleading for different reasons. These data could be used in simulation models to represent more general phenomena, but this type of computer modeling was still quite limited in the 1960s, and engineers did not put much stock in the results. In addition to concerns about the changing output of a vehicle with minute changes in driving conditions, there were also disagreements about the proper analytical instrumentation for measuring outputs. In 1966, there were competing papers at the SAE—some on the use of continuous mass spectrometry, others on gas chromatography—concerning the best techniques for analyzing nitric oxides and hydrocarbons. Engineers were not ignorant of contemporary developments in smog chemistry, but incorporating the cutting edge was difficult given the issues that remained unsettled among the chemists.[16]

TECHNOLOGICAL FIXES PART II: CHASING NITROGEN

By the early 1960s, research was focusing on additional pollution problems that were not addressed by reducing hydrocarbons. Cars were obviously not the only source of atmospheric hydrocarbons, and new research into smog chemistry brought a new concern into sharper focus: nitrogen oxides.[17] If nitrogen oxide could be reduced, the unburned hydrocarbons would have less to react with and would potentially form less ozone, another pollutant and lung irritant. However, an engine optimized to reduce hydrocarbons was also, in a way, an engine optimized to produce nitrogen oxides! While the problem with hydrocarbons focused on their disproportional prevalence in cold engines, nitrogen oxides were a product of hot engines. In fact, nitrogen oxides are formed through an endothermic reaction; they do not form without temperatures approaching 1600°C. At lower temperatures nitrogen and oxygen do not react. So the key to reducing nitrogen oxides was in reducing the temperature of combustion. Combustion temperatures could be reduced in warmed-up engines, but reducing combustion temperatures in a cold engine would lead to the car emitting more partially burned hydrocarbons, thus reversing the progress of a decade of hydrocarbon reduction efforts. The solution was the engine gas recirculation (EGR) system, which was one of the more finicky technologies produced to address emissions and one to which many consumers vocally objected.[18] Many autophiles objected to emissions control technologies on essentially libertarian, antiregulation grounds. Others felt that the EGR in particular compromised and limited the engine's performance. Still a third category disapproved of devices that weren't user-friendly to hobbyist mechanics. A fourth category was frustrated by finicky devices that needed frequent repair and made cars less reliable. Naturally, many users resisted on multiple grounds. Because it spawned a small red light on the instrument panel, indicating the need for service, many consumers focused their initial ire on the EGR as an exemplar of what was wrong with new required emissions control technologies.

EGR systems provided a way to recirculate a small amount of exhaust into the vehicle's intake manifold. Inert exhaust gas circulated back into the combustion system would reduce the temperature of combustion, basically, by reducing the oxygen available in the fuel-air mixture. EGR systems appeared in 1972 and were quickly adopted, despite consumers' complaints. General Motors in particular had trouble with customers' complaints about the EGR. Customers noted that their EGR warning light was lit up constantly, which led to some consumer skepticism about the system being a ruse for them to have more maintenance done on their cars. In an era of computer diagnostics, we take such situations for granted, but the EGR light was a very early step into a world in which expensive maintenance was often required on cars with no symptoms other than a dashboard warning light. EGR systems required frequent fine-tuning, since too much exhaust gas would make the engine run improperly and fail to reduce nitrogen oxides. EGR systems also required monitoring the engine's temperature at a number of locations, relays that opened and closed valves allowing recirculated gas in or not, and timers that prevented the EGR from engaging for about thirty seconds on an engine that was just being started. Following the successful introduction of EGR systems, nitrogen oxide limits could also be mandated: the Clean Air Act of 1970 mandated a 90 percent reduction in nitrogen oxides by 1976 and was frequently amended throughout the 1970s to ratchet up emissions standards.

By 1970 the SAE had turned considerable attention to the nitrogen oxides. Since the mid-1960s there had been at least fifteen papers on emissions at the annual SAE meeting in Detroit. The evolving interests of automotive engineers may be observed through an examination of these papers. For instance, papers prior to 1966 focused on the simpler problems of evaporation and measuring the effectiveness of PCV valves and other components. By the late 1960s engineers focused primarily on exhaust gases and communicated about various pollutants and how best to remove them. Once the key pollutants were identified, engineers thought of them as a hierarchy of challenges with carbon monoxide as the least problematic, unburned hydrocarbons as a bit more challenging due to their variety (over one hundred different hydrocarbons were in play), and the nitrogen oxides as the most troublesome.

TAKING STOCK OF THE FIELD OF EMISSIONS CONTROL IN 1970

By 1970, engineers engaged the full interaction and optimization problems of the three polluting compounds. California standards were moving to the national scene with the Clean Air Act of 1970 and the creation of the EPA to create, monitor, and enforce regulations and standards. In 1969, at the January SAE meeting, Charles Heinen of Chrysler presented a rather controversial paper titled "We've Done the Job—What's Next?" Heinen praised auto manufacturers for making exactly the needed improvements to clean up emissions and pronounced the job done. Clearly speaking for the automotive industry, Heinen also questioned the health effects of automobile emissions, writing, "automobile exhaust is not the health problem it has been made out to be."[19] He argued that since life expectancies for Los Angeles residents were equal to or greater than national averages, the concentration of exhaust and the prevalence of smog could not conclusively be called dangerous to health. Furthermore, he argued than any additional reductions in emissions would come only at a high cost—$10 billion nationally for new technologies and much, much more in decreased fuel economy for the user. Heinen was attempting to position the industry against the introduction of so-called catalytic afterburners aimed at eliminating even more CO, NO_x, and hydrocarbons. Catalytic afterburners would require tetra-ethyl lead to be removed from gasoline, since it would "clog" the catalysts. Heinen argued that this would be a considerable detriment to the operation of the automobile, because refineries had been adding lead to gasoline as an anti-knock agent since the 1920s. Lead-free gas would have to be reformulated to prevent engine knock through some other means. It is notable that several papers were given at the SAE between 1968 and 1975 on the options for lead removal from gasoline, most commonly by engineers at the oil companies, gasoline refineries, or chemical companies like Du Pont. These papers were much more sanguine about lead removal, pointing out that it would have health benefits that Heinen apparently overlooked.

Heinen's paper generated a rejoinder in a paper presented the following year by Phil Myers, a professor of mechanical engineering at the University of Wisconsin and president of the SAE in 1970–71. Myers's 1970 paper, "Automobile Emissions—A Study in Environmental Benefits versus Technological Causes," specifically cited Heinen but argued a different case. Perhaps reflecting his position in the academy versus Heinen's industry credentials, Myers argued, "we, as engineers concerned with technical and technological feasibilities and relative costs, have a special interest and role to play in the problem of air pollution."[20] He wrote that while analysis of the effects of various compounds wasn't conclusive from the perspectives of either atmospheric chemistry or public health studies, engineers had an obligation to make evaluations and judgments. To Myers, this was the crucial professional duty of the engineer, to collect data where possible but to realize that data would never override the engineer's responsibility to make judgments even when the empirical basis was thin. He strengthened this position in a paper he presented the following year on "Technological Morality and the Automotive Engineer," in a session he organized about bringing environmental ethics to engineers.

It is easy to dismiss Heinen's position as shilling for the automobile industry and to reward Myers for his ethically informed vision of the engineer's role in society. At stake between the two was the role of the engineer in society, especially with regard to making standards for the environment.

Both men credited the work already done, but Heinen took it as a job well done whereas Myers called it a process that should continue. At the end of his 1970 paper, Myers made his case for the role of engineers in both technological development and regulation, writing, "At this point, it would be simple for us as engineers to shrug our shoulders, argue that science is morally neutral and return to our computers. However, as stated by [O. M.] Solandt, 'even if science is morally neutral, it is then technology or the application of science that raises moral, social, and economic issues.'"[21] Myers then cited atmospheric chemist E. J. Cassell's proposal to control pollutants "to the greatest degree feasible, employing the maximum technological capabilities."[22] For Myers the beauty of Cassell's proposal was its nonfixed nature—standards would change (most likely become more stringent) as technologies were developed and their prices reduced through mass production. If acute pollution episodes threatened human well-being, then acceptable emissions standards could be raised with the awareness that pollution's social costs had also increased. For Myers, uncertainty constituted a call to action; to Heinen it was a call to inaction. Heinen saw fixed standards as both morally and economically superior, arguing, "we've done the job."

Myers also took a position with regard to the role of the engineer to spur consumer action. He wrote:

> Our industry recognizes that the wishes of the individual consumer are increasingly in conflict with the needs of society as a whole and that the design of our product is increasingly affected by this conflict. This is clearly the case with pollution—an individual consumer will not voluntarily pay extra for a car with emissions control even though the needs of society as a whole may be for increasingly stringent emissions control. . . . [T]here is universal agreement that at some time in the future, the growth of the automobile population will exceed the effect of present and proposed controls and that if no further action is taken mass rates of addition of pollutants will rise again.[23]

Myers was arguing that federal emissions standards should be written such that they could become more stringent in the near future. But Myers also believed that regulation was a necessary piece of this system, because it would serve to force the consumer's hand (or to put it another way, to force the market). According to Myers, engineers were the ideal mediators of these processes.

TECHNOLOGICAL FIXES PART III: THE CATALYTIC CONVERTER

The position regulators took to setting standards was somewhere between Heinen's hard standard and Myers's constantly and fluidly escalating ones. Fixed standards that changed frequently became the norm, but only after a disagreement about whether to measure the concentration of certain chemicals in the vehicle's exhaust or the mass of total exhaust-per-vehicle-mile. The latter was written into the Clean Air Act and remained the standard, but not without attack from those who argued that the exhaust concentration was a better proxy for clean emissions, since it allowed regulators to focus on various smog components. The early 1970s were really dominated by debates about how to modulate the remaining emission compounds catalytically in order to raise standards even higher. The first paper on catalytic devices appeared in 1973. The voices of European and Japanese engineers were prevalent in these sessions at SAE conferences and other meetings. The device that all these papers gestured to was the catalytic converter, which went by several different names before 1975, such as "catalytic afterburner" or "catalytic muffler."

The catalytic converter was developed as another technology to be added to the exhaust system.[24] Converters could catalyze a number of different reactions—converting carbon monoxide into carbon dioxide, oxidizing unburned hydrocarbons into carbon dioxide and water, and reducing nitrogen oxides to oxygen and nitrogen. The three-way catalytic converter, which stimulates all three reactions, was designed to be added to a vehicle with the TAC, AIR, EGR, and other emissions systems in order to produce a vehicle capable of meeting the requirements of 1976 to reduce hydrocarbons, carbon monoxide, and nitrogen oxides beyond the targets set by the Clean Air Act of 1970.[25] Automobile manufacturers also saw the catalytic converter as a tool to improve the performance of cars with emissions control devices, in response to customers' complaints and desires.

The challenge of catalytic convertors was threefold. First, they required that tetra-ethyl lead be eliminated from gasoline and replaced by new anti-knock agents. This challenge was met by 1972, in time for the introduction of the devices in 1975. Second, regulators forced the rapid introduction of devices, because emissions standards were changing often and quickly in the 1970s. Manufacturers felt that there was inadequate time to field-test them and assure their reliability and durability.[26] Consumers also demanded cars that got better mileage in response to the oil crises of 1973–74 and 1978. Yet the market was relatively flat due to the poor economy and automobile manufacturers were not flush with research dollars. This was a difficult period to be innovating. The third challenge was more complex. Catalytic convertors were parts of systems; they were not standalone add-ons. They required the use of other specific automotive technologies, and those technologies were completely untested.

Catalytic converters require oxygen sensors in order to manage combustion in the converter and prevent overheating, which could both damage the device and produce more nitrogen oxides. Oxygen sensors provide data to a logic gate, further increasing the demand for electronic circuitry in the automobile. Electronics in the automobile were not necessarily seen as a positive addition in the 1970s; at the time, consumer (nonmilitary grade) electronics, in general, were notoriously unreliable and expensive.[27] Systems like EGR, catalytic converters, and electronic fuel injection required a high level of data input and relay logic. These electronic requirements were similar to those required in antilock braking systems (ABS), electronic versions of which were introduced in 1978. This was a direction automakers feared their customers might reject. A lot was at stake surrounding decisions to computerize the automobile; systems such as electronic fuel injection and antilock braking systems took over a decade to find dominant market positions. In the end the introduction of electronics to manage engine combustion was an important watershed in automotive history but at the time automotive electronics seemed like a risky development to many manufacturers. Reinforcing the reluctance of automakers to introduce new technologies was the economic environment of the late 1970s and early 1980s.

The bumpy road leading to full computerization of the automobile by the 1990s opened a series of additional options for emissions control, which is a story unto itself. Complicated injection systems allowed engine efficiency and emissions to be improved. Gasoline direct injection (GDI) in particular allowed an extremely lean fuel-to-air ratio that reduced nitrogen oxides, but GDI is unworkable without a central processing unit in the engine to manage the injectors and valve timing in response to a series of nearly a dozen oxygen sensors, temperature sensors, and other data input generators. GDI also allows EGR to work more effectively and modulate combustion temperature. In addition, the computerization of the car led to a new measurement conception of emissions outputs. In 1990, federal legislation changed the way pollutants were measured, from parts per million in exhaust gases to grams per mile driven. Measuring these new outputs required different instrumentation and record-keeping by both auto manufacturers and local emissions inspection stations.

The transformation of the automotive emissions system through electronic control was part of a much broader process to digitally coordinate the functions of the automobile, from emissions to combustion to directional control. Microprocessors, developed outside the automobile industry in the 1970s, found multiple uses in the automobile by 1980.[28] The challenge to engineers was the coordination of different systems into a comprehensive adaptive control system. By 1990 many cars had central processing units, and this profoundly changed the automobile industry. Manufacturing an automobile required much higher-dimensional tolerances since nearly every mechanical part was being monitored and controlled by sensors. Maintaining automobiles would become a matter of plugging the car into a diagnostic computer to check all the sensors. Cars would need software updates as part of their regular maintenance, although these were usually invisible to the car owner. It was a new world. But one looming emissions concern remained—could the car (especially the SUV) become anything other than the bogeyman of a greenhouse-like atmosphere that threatened life on earth itself?

Myers's paper of 1970 raised the specter of carbon dioxide as a pollutant. He conveyed "increasing concern about [CO_2's] potential for modifying the energy balance of the earth, that is the greenhouse effect."[29] He called CO_2 the automobile's "most widely distributed and abundant pollutant." There is little evidence that many shared this concern, and to be fair, the issues of hydrocarbon, carbon monoxide, and NO_x emissions were more pressing at the time. CO_2 emission is unavoidably a product of combustion, and even the ideal combustion cycle converts hydrocarbons into CO_2 and water. There is no catalytic path to eliminating carbon dioxide after the combustion cycle; capturing it presents seemingly insurmountable problems, due to the volume of carbon produced and the obvious question of storage. The internal combustion engine has become a carbon dioxide–producing machine.

Over the past half century, the design of the automobile engine has changed significantly. Many of these changes have been in response to demands for engines that emit fewer smog-creating compounds, such as hydrocarbons, nitrogen oxides, and carbon monoxide. However, these developments have occurred against a backdrop of a constantly increasing number of vehicles on the road. It is honestly surprising that measurable gains have occurred at all, given the nearly sixfold increase in the number of vehicles in operation in the United States between 1950 and 2010, to say nothing of the increase in vehicles globally. In addition, the miles people drove those cars increased each year by 1.9 percent over the past sixty years.[30]

Also striking is the fact that the long-held assumption that carbon dioxide is harmless and therefore should be maximized has led to the car becoming a nearly perfect carbon dioxide–producing machine.

The technological fixes wielded by the automobile industry show both the promise and the peril of developing technologies to address problems created by other technologies. The global scale of automobile use renders the carbon dioxide problem produced by these cars so massive that it is impossible to solve. However, the case does point to the promise of solving technological problems in decade-long cascades (rather than "silver bullets") in conjunction with flexible legislation and regulation—and of the capacity of general-use technologies (semiconductors and computers, for example) to contribute in ways previously unthinkable.

A scientific consensus about air pollution was neither necessary nor sufficient for attacking the problem of automobile emissions. The technological fixes instituted in emissions control informed and interacted with, but did not determine, policy outcomes. There is obviously a lesson here for today's concerns about greenhouse gases and climate change. While policymakers tend to cling tightly to linear models of science begetting new technologies, in the case of automobile emissions the patterns seems to look more like technologies begetting more technologies, while opening up policy questions about how to put those technologies in consumers' hands. Furthermore, the removal of clearly dangerous emissions like carbon monoxide and nitrogen oxides was undoubtedly beneficial. While we marvel at the ways in which engineers have optimized carbon dioxide production in the automobile of the twenty-first century, we have also become skeptical about the possibility of future significant reductions of carbon dioxide emissions from the internal combustion engine. Fewer internal combustion engines and fewer miles driven are the only apparent ways out.

NOTES

1. Martin V. Melosi, "The Automobile and the Environment in American History," http://www.autolife.umd.umich.edu/Environment/E_Overview/E_Over view4.htm.

2. Jacobs and Kelly, *Smogtown*.

3. Zus Haagen-Smit, oral history taken by Shirley K. Cohen in 2000, http://oral histories.library.caltech.edu/42/01/OH_Haagen-Smit_Z.pdf.

4. Bonner, *Arie Jan Haagen-Smit*, 197.

5. Jacobs and Kelly, *Smogtown*, 70.

6. Haagen-Smit, oral history, 25.

7. Volti, "Reducing Automobile Emissions," 280–81.

8. Haagen-Smit, oral history, 26.

9. Hayes, *Motor Vehicle Emissions Control*, vol. 1.

10. Melosi, *Automobile and the Environment*.

11. Hayes, *Motor Vehicle Emissions Control*, vol. 2.

12. Hayes, *Motor Vehicle Emissions Control*, vol. 3.

13. Hayes, *Motor Vehicle Emissions Control*, vol. 4.

14. Caplan, "Smog Chemistry," 148.

15. Myers et al., "ABCs," 1678.

16. Brimblecombe, this volume.

17. Jacobs and Kelly, 125. See also Caplan, "Smog Chemistry"; Myers, "Automobile Emissions"; and Myers et al., "ABCs."

18. Hayes, *Motor Vehicle Emissions Control*, vol. 5.

19. Heinen, "We've Done the Job," 1952.

20. Myers, "Automobile Emissions," 658.

21. Ibid., 672.

22. Cassell, "Are We Ready," 799.

23. Myers, "Automobile Emissions," 672.

24. See Vinsel, "Federal Regulatory Management."

25. Hayes, *Motor Vehicle Emissions Control*, vol. 5.

26. Matsumoto, Matsumoto, and Goto, "Reliability Analysis," 728.

27. Johnson, "Unpacking Reliability."

28. See Johnson, *Hitting the Brakes*.

29. Myers, "Automobile Emissions," 660.

30. US Department of Energy, *Transportation Energy Data Book*, tables 8.1 and 8.2, http://cta.ornl.gov/data/chapter8.shtml.

REFERENCES

Bonner, James. *Arie Jan Haagen-Smit, 1900–1977: A Biographical Memoir*. Washington, DC: National Academy of Sciences, 1989.

Caplan, John. "Smog Chemistry Points the Way to Rational Vehicle Emissions Control." SAE paper at Annual Meeting, 1965.

Cassell, E. J. "Are We Ready for Ambient Air Quality Standards?" *Journal of the Air Pollution Control Association* 18 (1968): 795–803.

Hayes, B. D. *Motor Vehicle Emissions Control*. Vols. 1–5. Washington, DC: US Government Printing Office, 1978.

Heinen, Charles. "We've Done the Job—What's Next?" SAE paper 690539. Warrendale, PA: Society of Automotive Engineers, 1969.

Jacobs, Chip, and William J. Kelly. *Smogtown: The Lung-Burning History of Pollution in Los Angeles.* Woodstock: Overlook Press, 2008.

Johnson, Ann. *Hitting the Brakes: Engineering Design and the Production of Knowledge.* Durham, NC: Duke University Press, 2009.

———. "Unpacking Reliability: The Success of Robert Bosch, GmbH in Constructing of Antilock Braking Systems as Reliable Products." *History and Technology* 17 (2001): 249–70.

Matsumoto, K., T. Matsumoto, and Y. Goto. "Reliability Analysis of Catalytic Converter as an Automotive Emission Control System." SAE paper 750178. Warrendale, PA: Society of Automotive Engineers, 1975.

Myers, P. S. "Automobile Emissions—A Study in Environmental Benefits versus Technological Costs." SAE paper 700182. Warrendale, PA: Society of Automotive Engineers, 1970.

Myers, P. S., O. A. Uyehara, and H. K. Newhall. "The ABCs of Engine Exhaust Emissions," SAE paper 710481. Warrendale, PA: Society of Automotive Engineers, 1971.

Vinsel, Lee Jared. "Federal Regulatory Management of the Automobile in the United States, 1966–1988." PhD diss., Carnegie Mellon University, 2011.

Volti, Rudi. "Reducing Automobile Emissions in Southern California: The Dance of Public Policies and Technological Fixes." In *Inventing for the Environment,* edited by Arthur Molella and Joyce Bedi, 277–88. Cambridge, MA: MIT Press, 2003.

7

CHESS LESSONS
CONTROVERSY AND COMPROMISE IN THE MAKING OF THE EPA

Jongmin Lee

ON NOVEMBER 29, 2010, the Environmental Protection Agency administrator Lisa Jackson visited the Aspen Institute, a nonprofit organization, to begin a weeklong commemoration of the EPA's fortieth anniversary and celebration of its accomplishments on DDT, acid rain, recycling, unleaded gasoline, secondhand smoke, vehicle efficiency and emissions controls, environmental justice, toxic substances control, cleaner water, and public engagement. One of the EPA's first large-scale national projects had been the Community Health and Environmental Surveillance System, or CHESS, which attempted to combine epidemiological research and air quality monitoring. CHESS was an ambitious project, covering eight different geographic regions of the country and bringing together a wide variety of discipline and data analysis methods. The EPA intended to establish its federal authority over air pollution through CHESS, among other projects. However, CHESS proved extremely complicated to organize and implement and by the mid-1970s questions were raised about the validity of its data collection and analysis. CHESS became embroiled in a controversy of supposed data distortion and overinterpretation, resulting in a 1976 congressional hearing.[1] The EPA then used CHESS as a catalyst to reorganize its monitoring and research functions and reorient federal environmental management. Curiously, given its importance in the operations of the early EPA, there was no mention of CHESS's role in air pollution monitoring and health surveillance in the celebration of the EPA's fortieth anniversary. Moreover, interviews with former EPA professionals indicated that CHESS had no place in their collective memory. The absence of CHESS in history and memory provides a motivation to look for what it was and how it was forgotten.[2] This chapter explores the aims of the early EPA and the legacy of its now-forgotten flagship program, CHESS. The controversies surrounding CHESS make it representative of the kinds of environmental projects that have characterized the EPA since its inception—complex in its organization, ambitious in its scope, and conflicted in its outcome.

AMBITIONS

On January 1, 1970, President Richard Nixon urged the nation to pay attention to the environment in the new decade: "The 1970's absolutely must be the years when America pays its debt to the past by reclaiming the purity of its air, its waters, and our living environment. It is literally now or never."[3] Through public hearings and research reports, activists and scientists had shown that smog wasn't limited to Los Angeles and was becoming a national health concern. At the same time, local authorities received growing demands from constituents but had insufficient resources to meet them. Legislators responded by enacting and amending the Clean Air Act and the Clean Water Act, but federal efforts were limited in scope and scale.[4] To make things worse, some military bases and federal research centers were not complying with the very statutes the federal government wanted to impose on the states. Recurring pollution episodes alerted activists who, in turn, mobilized a concerned public. On July 9 Nixon responded with Reorganization Plan No. 3, creating the Environmental Protection Agency, a federal organization designed to monitor and protect air, water, and land wherever pollution occurred.

What did it mean to have an integrated system of pollution control at the federal level? Why did the federal government need to protect the environment? What impacts did the EPA's activities have on how scientists, engineers, and medical professionals understood the problem of air pollution, one of the leading sources of public concern? Many health professionals linked air pollution to asthma or irritation of the eyes or nose; however, a lack of detailed indices and measurement devices for pollutants hindered accumulation of comparable data. As was the case in earlier eras, health effects researchers were still struggling to identify causal and proportional relationships between pollutants and disease, and knowledge of the transport of pollutants was insufficient to set up effective strategies to remediate large-scale pollution.[5]

In order to respond to urgent needs for environmental protection, EPA managers swiftly created an organizational structure to address its myriad challenges. The first leader of the EPA, William Ruckelshaus, set up ten regional offices to cooperate with state and local governments. The agency assumed responsibilities from an alphabet soup of predecessors and was also assigned new responsibilities that no agency had overseen. The EPA had to integrate research on different types of pollution (e.g., air, water, chemicals, and solid waste), construct environmental standards, and enforce regulations.[6] In doing so, it depended on reliable and timely information on a vast range of potential harms. Like other new regulatory and research agencies of the early 1970s, the EPA also had to establish its authority. In short, the new agency faced several organizational and epistemic challenges in defining a coherent agenda and delineating the boundaries of government research and regulation. All of this would have a distinct impact on how the EPA approached the question of what the federal responsibility for air was and how air quality could be monitored and managed.

The story of CHESS shows how EPA scientists and engineers tried to integrate human and environmental health by building systems to deal with new geographical, bureaucratic, and epistemic challenges. This chapter documents how the EPA instituted its research, monitoring, and regulatory practices and reveals how those practices redefined the environment that the EPA was established to protect, govern, and improve. I apply the notion of coproduction—showing how engineering practices were incorporated into regulation and organizational change, which in turn influenced technical knowledge and identity formation—to explain how technical and social orders are created, stabilized, and altered through constant interaction.[7] I also build on "regulatory science" approaches to describe and analyze techno-scientific activities originating within the EPA.[8] Doing so reveals the changing role of pollution monitoring, the asymmetrical relationship between research and regulation, and a subsequent policy priority shift from community health assessment to pollution-control technology within the EPA. Ultimately, it reveals how the quest to exert federal authority over air quality in the United States played a part in shaping the nation's system of environmental protection.

From its inception, the EPA took on two distinct roles: (1) research manager for the nascent fields of environmental science and engineering, and (2) primary regulator for all the environmental pollution in the United States. The organizational structure designed to fulfill these two missions generated two different kinds of departments within the agency. First, there were offices on air, water, pesticides, radiation, and solid wastes that were based on categories of pollutants or relevant media. In addition, there were three more offices that cut across different pollutant and media types and were instead based on their functions within the whole agency. These were the Office of Research and Monitoring, the Office of Planning and Management, and the Office of Enforcement and General Counsel. Research and monitoring, as functions, subsumed about one-third of the EPA's budget and human resources. In 1971, EPA's budget for internal research was $125 million, with nearly 1,900 scientists, engineers, and technicians working to specify the pollution thresholds and find the best available control technologies to keep pollutants below those thresholds.[9]

Scientists and engineers at the EPA wanted to draw from off-the-shelf methods and instruments as well as identify areas where technological innovations could be developed to achieve a position of leadership in environmental monitoring. They measured pollution in water, air, and soil using proven methods and instruments from sanitary, mechanical, and chemical engineering. They also performed in-depth research on the composition and transport of pollutants and applied new insights from ecology and systems theory to attempt to model interactions of different types of pollution.[10] Monitoring and evaluation of remote sites required technical innovations involving the networking of command, control, and communication systems to identify, analyze, and characterize pollutants and their interactions.[11] Ruckelshaus argued that by doing so, the EPA would be able to "take remedial action before a problem became too intractable."[12]

Achieving national goals demanded new organizational structures. The Office of Research and Monitoring's first director, Stanley Greenfield, wanted to position the EPA as a leader in environmental monitoring, setting the standard for efforts by other federal agencies.[13] Greenfield organized the existing local and regional laboratories, monitoring sites, and field stations into a system of four thematic National Environmental Research Centers (NERCs) located in Cincinnati, Ohio (on water pollution control); Corvallis, Oregon (on ecological effects); Las Vegas, Nevada (on radiation monitoring); and Research Triangle Park, North Carolina (on air quality and toxicology).[14] The location of the NERCs involved considerations of resources, politics, and the institutional continuity of preceding agencies.[15]

On the regulatory side, the EPA was a high-profile and politically contentious agency that actively managed its public image to establish credibility. Ruckelshaus recalled, "It was important for us to advocate strong environmental compliance, back it up, and *do* it; to actually show we were willing to take on the large institutions in the society which hadn't been paying much attention to the environment." In 1972 the EPA initiated fifteen legal actions every month against a variety of institutions from municipalities to major industries.[16]

CHESS: BUILDING A PROTECTIVE SYSTEM

A collective, longitudinal epidemiological study of selected geographical areas had begun under the National Air Pollution Control Administration in 1968 in the name of the Health Surveillance Network. The areas covered were Los Angeles, California; Salt Lake Basin, Utah; St. Louis, Missouri; Chattanooga, Tennessee; Birmingham, Alabama; Charlotte, North Carolina; New York, New York; and the New York-New Jersey border. Not without controversy, the EPA took over the existing Health Surveillance Network in 1970 and expanded its portfolio of projects, giving the study a new acronym—CHESS—and locating its headquarters with the NERC in Research Triangle Park (RTP).

The chief of the Office of Research and Monitoring (ORM), Stanley Greenfield, intended CHESS to be a showcase combining exemplary research on pollution monitoring and epidemiology in order to demonstrate the value of managing air quality to Congress, industry, and the public.[17] It was intended to show the benefits of combining environmental monitoring and health protection, but these benefits proved difficult to assess. Evaluation of existing air quality standards was CHESS's first goal, with an aim to demonstrate their benefits. To implement the timely enforcement required by the Clean Air Act of 1970, the EPA promulgated the National Ambient Air Quality Standards for carbon monoxide (CO), hydrocarbon (HC), ozone (O_3), particulates (PM), sulfur dioxide (SO_2), and nitrogen oxides (NO_x) in 1971.[18] The act required that the EPA use the latest scientific information to set various standards for public health, welfare, and short- and long-term exposure.[19] Setting national standards, however, proved to be a daunting task. Air pollution research done in other nations or by international organizations provided broad guidelines, but domestic research supporting certain concentration thresholds was needed to satisfy a diversity of interested parties, including industrialists, environmentalists, politicians, and scientists.[20] Such consensus was elusive.

Integrating various existing projects into a new program was also a major hurdle, in part because of the daunting complexity and in part due to changes in personnel. Greenfield assigned responsibility for CHESS to NERC–RTP under the directorship of Delbert Barth, who had previously been in charge of air pollution science at RTP. Earlier in his career Barth worked as a radiation health specialist in the US Army and the US Public Health Service and then became head of the Bureau of Criteria and Standards of the National Air Pollution Control Administration. His tenure was short-lived, however, and in 1972 he was reassigned to NERC–Las Vegas because of his background in radiation safety.[21] The second director of NERC–RTP, John Finklea, was a physician and public health administrator who had worked for the National Air Pollution Control Administration and had previously been the director of the expanded Health Effects Research Laboratory at the EPA.[22] Finklea redesigned CHESS as a systematic study focused on three types of research: aerometry, epidemiology, and statistical correlation of health effects with atmospheric pollution.

CHESS defined "community" as a middle-class, residential segment of an area, containing three or four elementary schools and sometimes a secondary school.[23] The EPA justified the selection of middle-class neighborhoods by claiming that these communities are "migrationally stable." Comparing data from one region to another required accounting for demographic differences (e.g., age, sex, race) and exposure information (e.g., diet, water, smoking status, daily movement). Each community was chosen to represent an exposure gradient for certain pollutants; otherwise, communities were similar in meteorological and socioeconomic conditions. The gradients were, for example, PM gradient with low SO_2 for St. Louis and Birmingham/Charlotte, and SO_2 gradient with low PM for Salt Lake Basin (see fig. 7.1).[24]

CHESS researchers attempted to quantify pollutant burdens in the exposed population and focused on the respiratory "health" of the subsets of the population vulnerable to pollution. While there was general agreement on the connection between pollutants and respiratory diseases, scholars still debated how much of an influence came from each pollutant.[25] That was why CHESS employed several gradients of SO_2, PM, O_3, and NO_x. In addition, scientists were still searching for other possible explanations of certain diseases. For example, researchers considered asthma to be aggra-

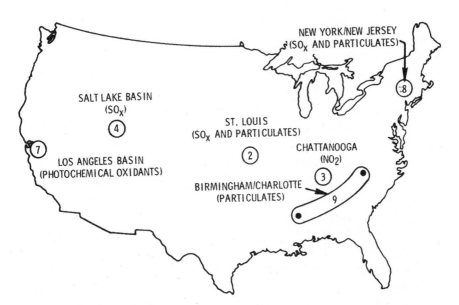

FIG. 7.1. CHESS Areas as of March 1973. SOURCE: EPA, *Health Consequences of Sulfur Oxides*, 1–4. Each number inside the area's circle shows the number of the communities included. Note that Los Angeles is misplaced on the map.

vated by environmental exposure to pollutants, but they also recognized other causes such as allergies, patients' immune systems, and the weather.[26] Health effects research required evidence from several studies as well as multiple approaches, including toxicology, clinical research, and epidemiology. As each approach had its own strengths and limits, laboratories employed all of these approaches to complement each other.

Epidemiological studies measured the relatively long-term effects of living in a polluted area. This approach was valuable because it facilitated an exploration of real people exposed to the variety and complexity of a real environment. However, epidemiological studies had several drawbacks. Epidemiological studies were time-consuming and expensive. They required managing a trained workforce of field-level researchers with sophisticated communication and interviewing skills. Epidemiological studies also depended on statistical research involving different kinds of data that had to be managed and made coherent. Furthermore, even carefully planned studies were unable to take account of every social and environmental variable that affects human health. Therefore, when an association was inferred between a particular exposure and an effect, toxicologists or clinical researchers still needed to corroborate this association with further studies.[27] Thus, renowned epidemiologists such as David Rall and Ian Higgins stressed that further studies would decrease the margin of error and result in more accurate standards.[28]

The design of health indicators in CHESS showed researchers' emphasis on showing subtle associations rather than strong causation. The chosen indicators reflected morbidity rates (the occurrence of diseases) more than mortality rates (the occurrence of death). The indicators facilitated exploration of low-dose effects on health and particularly focused on children, the elderly, and asthmatics. By targeting vulnerable communities, CHESS emphasized its relevance for the populations most threatened by pollution. Thus, to fulfill the EPA's mandate, CHESS researchers looked at correlations between pollutant exposures and respiratory health concerns of the subgroups within the communities who most needed attention.[29]

For CHESS the "environment" meant the ambient air in the study communities. CHESS's monitoring activities were different from its predecessors in two ways: standardization and integration. CHESS's headquarters in North Carolina was responsible for providing compatible sampling devices to each community. Proper calibration of devices was essential to obtain reliable and manageable data. CHESS researchers used pollution monitoring guidelines to provide contractors with standardized procedures.[30] For each community, the EPA built air-monitoring shelters and installed particulate samplers, NO_2 bubblers, and analyzers.[31] The EPA chose its contractors from a pool of academic or private researchers, who were to be responsible for installing and checking the bubblers and filters and returning them to CHESS headquarters for calibration. Contractors also gathered meteorological data such as wind speed and direction, temperature, and relative humidity.[32]

CHESS's "surveillance" function aimed to document the health benefits of pollution control, one of the study's ultimate goals. Researchers surveyed communities' health through questionnaires, phone interviews, diaries, and function tests. They either distributed one-time questionnaires about acute and chronic respiratory disease to families through schools or conducted biweekly telephone interviews. Pulmonary function testing measured the lung performance of children (see fig. 7.2). The asthma panel studies recruited participants and requested them to report their asthma symptoms by mailing back diaries. After each round of study, these data were coded into the optical scan sheets for data processing at the Research Triangle Park facility.

The ambitious goals of CHESS could not be realized unless the two "sys-

FIG. 7.2. Pulmonary function testing of California schoolchildren in 1973.
SOURCE: Statewide Air Pollution Research Center, *CHESS California Progress Report*, 91.

tems" of pollution and health were connected. In 1971, Greenfield announced to fellow meteorologists his "commitment to the total system concept, the inter-relation of environmental problems."[33] CHESS's system concept had several roots ranging from systems engineering to ecosystems theory. In this regard, CHESS was itself a giant supersystem of research and monitoring that aimed to connect two previously independent systems: pollution monitoring and health surveillance. In an attempt to facilitate the operation of CHESS, researchers employed logistical networks and data processing methods to correlate air pollution with health. Researchers faced several challenges in combining preexisting frames for thinking about the environment into a systems framework.

The EPA's pollution monitoring system emulated the operations research approach of military-industrial projects.[34] The MITRE Corporation's report on the National Air Pollution Control Administration's contract research shows a schematic flowchart of a nascent pollution monitoring system.[35] First, the headquarters standardized the manual of exemplary stations and disseminated it to the monitoring branches. Using a detailed timeline, headquarters assigned responsibilities and delegated authority to the branches.

By setting up headquarters, branches, and communications between them, the EPA developed CHESS's monitoring network to be an infrastructure for further environmental management. The reliability of this monitoring network depended on each branch's capacity to collect data and the headquarters' capacity to process and analyze it. CHESS administrators focused on the flow of information and mechanisms for generating timely responses at the expense of assuring the quality of data. This decision foreshadowed CHESS's downfall.

Researchers understood CHESS as a data processing system that connected health and pollution systems. Through bioenvironmental measurements, researchers codified the health and pollution information of the communities. They processed and analyzed the data seeking correlations between health conditions and air pollution gradients. After an initial test with certain sets of data and feedback, researchers corrected the methods to enhance correspondence between measurements, data collection, and information synthesis in the system. CHESS, however, faced several challenges to achieving its objectives. As the research spanned several years in various locations, it was crucial for researchers to update each other on their progress. Quality assurance, the maintenance of standards for both measurements and data processing, became a recurring problem. Device calibration and personnel training were perpetually underfunded.[36] Furthermore, CHESS publications were continually delayed, because incoming data were much more extensive than expected and data processing became more time-consuming. At the same time, the conversion from an IBM 360–50 to a UNIVAC 1110 required recoding data into a new computer system.[37] Managers underestimated the challenge of conversion and the additional resources required, which delayed data processing and led to the accumulation of unprocessed CHESS data.[38] In light of their ambitious goals for a new system of environmental protection, CHESS practitioners were not successful in controlling the quality of their data processing practices.

In the final analysis, coordination of various systems was the critical challenge CHESS faced on many different levels. CHESS's research design seemed flexible enough to link together and implement pollution monitoring and health surveillance systems as they evolved over the years. Each area study was independent but also connected with other areas' results to form a comprehensive survey. However, intractable problems emerged as the project progressed. Contractors sometimes demanded more time and budget than expected. Earning and maintaining the trust of research subjects required researchers' attention from the field to headquarters. Databases were still under development and required more work and resources than expected. Conversion to a new computer system delayed the processing and validation of data, which were already behind schedule. These logistical challenges ultimately threatened the existence of CHESS itself.

CONTROVERSIES AND CHALLENGES

In the early 1970s, engineers and scientists at the EPA tried to establish their credibility as guardians of the environment, with CHESS as their marquee example. But CHESS soon became embroiled in a controversy that altered the relationship between research and monitoring in the EPA. Both in-house and external reviewers expressed concerns and reservations about CHESS's measurement techniques and health survey and analysis methods. A 1973 National Academy of Sciences report concluded that "the need for a great deal of information in the shortest possible time has meant that the EPA has been forced to attempt too much too superficially."[39] Authors of the 1973 report were straightforward in pinpointing technical faults and conceptual limitations, but were cautious in questioning CHESS researchers' capacity or integrity. The report of the 1974 Rall Committee on the Health Effects of Sulfur Oxides argued for more research on synergistic effects of pollutants, more epidemiological data, and a better research database. The Rall Committee suggested assembling a group of outside reviewers for continuing advice.[40] A 1975 National Research Council report, although largely supportive of CHESS, noted "a number of experimental shortcomings."[41]

The epidemiological elements of CHESS attracted the most criticism. Finklea and his CHESS colleagues published a monograph, *Health Consequences of Sulfur Oxides*, in 1974, which was a collection of twenty papers based on the 1970–71 school year data.[42] Before its publication, the authors received comments on the 1972 draft of the CHESS monograph from colleagues in and out of the agency. Critics argued that the authors were overinterpreting their data, although the authors didn't agree. The EPA's Science Advisory Board (SAB), a public advisory group providing extramural scientific advice, convened a review panel of five specialists, including two statisticians, an epidemiologist, a toxicologist, and an environmental health engineer.[43]

In a report published in March 1975, the SAB panel criticized CHESS's epidemiological methods. They argued that CHESS limited its sample to middle-class and suburban populations, sampling was not done on a random basis, and response rates varied. The panel also pointed out ambiguous definitions of respiratory disease-states and symptom severities, which caused difficulties in comparing results. Panelists then recommended that EPA employees publish their results more in peer-reviewed scientific journals and collaborate with academic statisticians.[44] The SAB criticized specific methods of CHESS, but it also supported EPA's technical assumptions and regulatory goals. The SAB report argued that, once these weaknesses were corrected, CHESS would contribute significantly to the health information regarding atmospheric pollutants. In its conclusion, the SAB panel recommended the continuation of epidemiological studies as vital to the "understanding of the effects of air pollution on health."[45]

The SAB report was critical of the epidemiology practiced at CHESS. When collecting health information, they mostly depended on questionnaires and interviews, the results of which varied idiosyncratically from subject to subject and depended on interviewers' skills. Public reports about pollution colored participants' responses and attitudes, and in some cases respondents self-reported medical symptoms, such as asthma, without evaluation by physicians, making the results vulnerable to scientific criticism.

It did not take long for criticisms to expand to the aerometry section. Throughout the execution of CHESS, infrequent calibration of instruments used in different communities made it difficult to compare results. For example, Anthony V. Colucci, a former EPA employee turned consultant, analyzed CHESS aerometric samples and argued that the filters used for measuring sulfates actually contained varying amounts of sulfates from the manufacturing process.[46] Furthermore, some contractors were careless about checking and calibrating instruments, which then yielded less reliable data. Quality assurance was slowly introduced, but the inconsistent previous data from various CHESS areas hindered integrated analysis of the pollution–health relationship.[47]

Questions about the validity of CHESS data created conflicts between industry and the EPA. The coal and utilities industries, seeing the CHESS controversy as a chance to call for a moratorium on clean air legislation, lobbied to request amendments to the Clean Air Act.[48] Greenfield, who had left the EPA to become a consultant in California, expressed a critical but equivocal position. Greenfield's work was funded by the Electric Power Research Institute (EPRI), which was closely related to the electric utility industry. He based his critique of the CHESS report on his own analysis of the raw data, but he still endorsed and praised Finklea.[49]

The EPA, with the help of the scientific and engineering communities and the Natural Resources Defense Council, argued that CHESS data were still a meaningful contribution to pollution and health research, but opinions were mixed. Carl Shy, a former EPA epidemiologist and professor at the University of North Carolina, assumed responsibility for analyzing the rest of the data, while H. Daniel Roth, a former EPA statistician and consultant sponsored by Electric Power Resource Institute, argued that the remaining data on pollution should be abandoned.[50] In response to critiques of the validity of its data, EPA restricted use of the not-yet-analyzed CHESS data for policy, and no more CHESS reports were officially produced.[51] Epidemiologists and economists, however, continued to discuss the methodology and evidence of CHESS.[52]

In 1976 the *Los Angeles Times* reported that former CHESS head John Finklea, by then the director of the National Institute for Occupational Safety and Health (NIOSH), had deliberately distorted the CHESS findings. Reporter William B. Rood accused Finklea of rewriting the work of agency scientists, deleting important qualifiers, and said he "overrode agency scientists' objections to publishing estimates of the health impact of pollution which were either statistically dubious or unsupportable." Rood also argued that by "relying heavily on the disputed CHESS studies, [the] EPA [had] called for controls on sulfur pollution that would cost power companies and ultimately American consumers billions of dollars."[53]

Such media attention triggered an agency investigation, followed by a congressional hearing in April 1976. Contrary to Rood's assertion, congressional staffers had a difficult time finding colleagues critical of Finklea. EPA administrator Russell Train and researchers were invited to testify about Finklea and CHESS. There were also outside commentators, some of whom had previously worked with Finklea in and out of the EPA. *Science* reporter Philip Boffey summarized witnesses' statements that the revisions made were to establish uniformity and completeness and Finklea had indeed left out qualifying statements, but also that the errors were never systematic or intentional. Although short deadlines had led to compromises, Boffey added, Finklea's draft had gone through review, criticism, and revision, and the final document was a valuable piece of work. The document employed "best judgment" approaches, and not "worst case" analyses.[54] Robert Buechley, an epidemiologist at the University of New Mexico and a former analyst for the National Air Pollution Control Administration and the EPA, was the most critical of Finklea. However, Buechley acknowledged that he did not actually work on CHESS with Finklea and was only repeating other colleagues' complaints.[55] The outcome of the hearing was to require more stringent oversight of the EPA's research and ultimately led to the separation of EPA's research function from its regulatory role.

In sum, criticism of CHESS moved from questions about its research design and execution to challenges of the researchers' personal integrity. These controversies led to restructuring the EPA's research and regulatory functions. During heated discussions about CHESS and subsequent delays of the air pollution legislation schedule, the EPA's Office of Research and Development (ORD), the successor to the Office of Research and Monitoring, abolished the NERC system in 1974 and reorganized its offices and laboratories. Washington headquarters now directly coordinated four new clusters of ORD laboratories. The new structure emphasized the development of pollution control technologies and quality assurance of data.

Congress responded to the reorganization by passing a new law specifying the EPA's role and responsibility in environmental research and development. In 1976, the House Committee on Science and Technology published an investigative report saying that the EPA's planning and management of overall research and development should be reformed. In 1977, Congress passed the Environmental Research, Development, and Demonstration Authorization Act, which directed the EPA administrator to submit an annual five-year plan for research, development, with commentary from its Science Advisory Board. It further directed that the advisory board independently report to Congress on the findings and recommendations of the 1976 CHESS investigative report.[56]

As the EPA reoriented its priorities in consultation with Congress, monitoring became less visible, first in practice and then in organization. No new monitoring projects were initiated, unless they were in support of urgent enforcement cases or long-term research. Long-range, exploratory research was separated from day-to-day monitoring activities and drew more attention and money. Monitoring of air pollutants became routine work and was downgraded to be supportive of other research and regulatory actions. The Office of Monitoring was combined with the Office of Environmental Engineering to take only "short-term activities like quality assurance, monitoring, and analytic responses to the immediate needs of other agency programs."[57]

Ironically, the CHESS controversy generated more impetus for epidemiological studies, despite those studies being downsized at the EPA. Inspired by the broad agenda of CHESS, and challenged by the attack on epidemiological research in general, James L. Whittenberger, an epidemiologist who was the Science Advisory Board review committee chairperson in 1975, persuaded his Harvard colleagues to initiate the "Harvard Six Cities Study." As a result, CHESS inspired other epidemiologists to conduct research on indoor air quality and to examine the relationship between particulates and cardiovascular mortality.[58] In sum, although CHESS ceased to be a pioneering project, its components of monitoring and epidemiology became standardized and evolved into stronger and more robust endeavors.

As monitoring became more routine and less visible, industrial research communities aimed for new goals. For industry, this change in monitoring meant tactical success. The downfall of "command and control" regulation opened up spaces for new regulation through economic incentives.[59] The CHESS controversy had actually delayed legislative efforts to update the criteria standards. For scientific communities outside of industry, downsized monitoring opened up new opportunities to pursue long-term research. Philip Handler, the president of the National Academy of Sciences, supported the Environmental Research, Development, and Demonstration Authorization Act by creating a cadre of academic scientists who appealed to the EPA's Office of Research and Development to sponsor long-term, often basic, research.

CHESS's Legacy in Health and Pollution Research

CHESS and its subsequent controversy represented an important moment for the coproduction of knowledge about air pollution and disease and federal environmental management in the 1970s. One might even say that the meaning of "toxic airs" was redefined by EPA's efforts and federal jurisdiction over pollution. The EPA's research agenda and mechanisms for regulation were coproduced and together shaped the EPA's authority and identity. Only through organizational and research trial and error, sometimes painfully on display in the public realm, was the EPA able to attempt to define what could be regulated and the public health benefits of doing so. With a mandate to deal with environmental degradation, the EPA relied on various measures, including chemical monitoring, health research, enforcement, and engineering. The practices of planning, executing, and evaluating these measures were dependent on organizing economic and human resources (and shortcomings) as well as leveraging the EPA's authority in Washington politics and state bureaucracies. Studying CHESS showcases the dynamic relationships among monitoring, technical innovations, and organizational priorities in the new system of environmental protection.

Because of the controversies surrounding CHESS, it may be useful to ask whether CHESS failed to perform the functions for which it was designed. The complexity of motives for CHESS makes this a difficult question to answer. One complexity stems from the organizational structure of CHESS, which, by intent, shaped knowledge and influence flows asymmetrically from research to regulation. William Ruckelshaus supported Stanley Greenfield's idea of establishing the ORM independently from the program offices on air, water, pesticides, and solid waste. Greenfield then proposed the NERCs as institutions that could focus on particular environmental concerns, such as water and air pollution, radiation, and ecological effects—a move that provided a sense of continuity with the pre-EPA National Air Pollution Control Administration. Greenfield believed that scientific researchers were most productive when they were kept insulated from economic reality, regulatory needs, and influences inside and out. So influence and information flowed more freely from the crosscutting offices to the program offices; however, the EPA expected regulators to be aware of current research outcomes, even if researchers weren't always on top of current regulatory goals. This asymmetry reversed after the CHESS controversy. Regulators grew skeptical of relying on the latest research results from the ORD, but researchers were expected to know the EPA's regulatory timeline and were mandated to provide an annual five-year research plan to Congress.

CHESS was also motivated by the goal of developing new methodologies for research and monitoring. Researchers and managers were proud of their progress in air pollution research, and they wanted to expand the result into other media. By designing CHESS area studies with diverse exposure gradients, Finklea built on accomplishments from earlier pollution control measures but also emphasized the remaining issues that needed urgent action. One facet on which to judge CHESS is the researchers' success in technological and methodological innovation. However, again the project's record is mixed and the controversy about the validity of its data highlights the problem with innovative methods—new methods and instruments are particularly vulnerable to both scientific and political challenges, as was the case with CHESS.

Additionally, Finklea wanted CHESS to show how monitoring and research mutually benefited each other. For example, aerometry missions contributed to the development of new pollutant measurement techniques, and epidemiological research generated health indicators to measure subtle information about morbidity. CHESS researchers developed more advanced strategies to connect pollution monitoring and health surveillance systems and introduced, tested, and revised new ecological concepts and data processing techniques. The proliferation of novelty led to confusion about standardizing instruments and data collection techniques in the field and led to controversies over data validation.

As CHESS's style of integrated pollution monitoring and health research was downsized in response to allegations, the EPA's regulatory focus shifted from a community health approach to a technology-based one. Academic, bureaucratic, and corporate resistance to the community health and environmental surveillance elements of CHESS opened up a space for an alternate "control technology" approach to expand inside the EPA. Control technology proponents argued for using available technologies in order to bring about quick reductions. For example, catalytic converters and scrubbers became de facto standards for subsequent mobile and stationary air pollution regulations, respectively.[60] Reorganization and rearrangement of the Office of Research and Development also reflected the upscaling of energy-related development.[61] Subsequent shifts in regulation from "command and control" to performance-based economic measures eventually influenced the category and definition of "the environment." In other words, the EPA's reorganization following CHESS not only reflected the shifting priority of environmental policy but also fortified it by changing the focus of monitoring and regulatory practices.

CHESS's most lasting impact came from the movement of its personnel out of the EPA. As monitoring became routine and the relative emphasis on research and regulation reversed, Finklea and colleagues moved to NIOSH, where they set up criteria standards, designed surveillance networks, and again faced corporate resistance to occupational health and safety measures. Greenfield and colleagues moved to California to become consultants for the electric power industry and municipal and state governments. Barth, Shy, and colleagues moved into academia to produce the next generation of regulatory professionals.

In the 1970s EPA engineers, scientists, and managers created and transformed the US system of environmental protection and redefined the meaning of the environment on a national scale. By looking at CHESS and its monitoring, technology, and organization against the background of the coproduction of clean air, healthy communities, and efficient regulations, I offer a first step in tracing controversies and compromises that shaped the protective system we live in. In the end CHESS's legacy was buried by the controversy generated over its organizational structure, data collection techniques, and data analysis methods. CHESS's legacy lives on only through the future work of CHESS veterans. However, as a project that may have been too novel and ambitious to succeed, it played an important role in the future of environmental regulation by making it clear that overreach in both research and regulation would have negative consequences for the authority of the EPA. After CHESS, projects were more modest about their goals, less novel in terms of their monitoring technology, less interdisciplinary, better regulated, and more clearly contained with a single organizational unit.

NOTES

1. Rood, "EPA Study."

2. John Bachmann was the only exception. His very cautious reply to my question about CHESS was based on his 2007 article, from which he quoted word for word. Bachmann, interview by the author, Mar. 14, 2012.

3. Richard Nixon, "Statement About the National Environmental Policy Act of 1969," Jan. 1, 1970, online by Gerhard Peters and John T. Woolley, *The American Presidency Project*, accessed Feb. 4, 2013, http://www.presidency.ucsb.edu/ws/?pid=2557.

4. On the Clean Air Act, see Polli, this volume.

5. See Hamlin, this volume.

6. US Environmental Protection Agency, *Progress Report*.

7. Jasanoff, "Ordering Knowledge," 13–45; Latour, *Pasteurization of France*.

8. Jasanoff, *Fifth Branch*, 61–83; Irwin et al., "Regulatory Science," 17–31; Dear and Jasanoff, "Dismantling Boundaries," 759–74; Swedlow, "Cultural Coproduction," 151–79; Lee, "Engineering the Environment."

9. EPA, *Directory*; EPA, *Progress Report*; Greenfield, "Overview," 27.

10. EPA, *EPA's Monitoring Programs*.

11. Shaller, *System Development Plan*.

12. Ruckelshaus, "Conquest," 3.

13. "ES&T Interview," 990. Schooley, *Responding to National Needs*, 340, 402–20.

14. John M. Moore, "EPA Las Vegas Laboratory Early History, 1952–1970," EPA Historical Document Collection, Box 7 on EPA Organizational History, Washington, DC; "Chronological History of EMSL-Las Vegas," EPA Historical Document Collection, Box 7 on EPA Organizational History, Washington, DC; EPA, *National Environmental Research Centers*; EPA, *EPA at Research Triangle Park*.

15. "ES&T Interview," 990–92; EPA, *Progress Report*; EPA, *National Environmental Research Centers*, 1.

16. US Environmental Protection Agency, William D. Ruckelshaus oral history interview, interviewed by Michael Gorn, EPA 202-K-92–0003, US EPA History Program, Nov. 1992, accessed Feb. 4, 2013, http://www.epa.gov/aboutepa/history/publications/print/ruck.html. Emphasis is from the text.

17. National Industrial Pollution Control Council, *Air Pollution*, 9.

18. Standards for these pollutants since have been continually revised, some dismissed, and new standards added.

19. Clean Air Act of 1970, 42 USC.

20. See Rothschild, this volume.

21. Barth, interview by the author, Feb. 10, 2011; "Biography of Delbert S. Barth, Assistant Surgeon General, Rear Admiral, USPHS (ret.)," July 7, 2008, http://usphsengineers.org/index.php/history_of_phs_engineers/109-delbert_barth.html.

22. Kilburn, "In Memoriam," 293.

23. Riggan et al., "CHESS," 111–23.

24. US Congress, *Conduct of EPA's CHESS Studies*; Riggan et al., "CHESS," 111–23.

25. Heimann, *Air Pollution and Respiratory Disease.*

26. Boushey and Fahy, "Basic Mechanisms of Asthma," 229–33.

27. Corman, *Air Pollution Primer,* 57.

28. US Congress, *Conduct of EPA's CHESS Studies,* 97–100; Rall, "Review of the Health Effects."

29. Finklea et al., "Health Intelligence for Environmental Protection," 101–109.

30. The government of the modern state is interested in the monitoring of its resources. Monitoring of its people and the environment, in return, contributed to the formation of the modern government. To understand how the state makes those resources visible and simplified, see Scott, *Seeing Like a State.* On the efforts to quantify nature and society and to build a trust around it, see also Porter, *Trust in Numbers.* On the standardization of instruments and its influence in science and industry, see Joerges and Shinn, *Instrumentation between Science, State, and Industry.*

31. American Institute of Chemical Engineers, *Pollution and Environmental Health.*

32. Barnard, *CHESS Air Pollution Monitoring Handbook*; Statewide Air Pollution Research Center, *CHESS Studies California Progress Report,* 7–8.

33. Greenfield, "Our Future Is Up in the Air," 846.

34. Shaller, *System Development Plan.*

35. Ibid.

36. Bregman Co., *Technical Primer.*

37. Kelsey, Lowrimore, and Smith, "Conversion of CHESS," 282–88.

38. Greenstein, "Lock-In," 258–61.

39. Cited in EPA, *Addendum,* 119.

40. Rall, "Review of the Health Effects," 97–121.

41. US National Research Council, *Air Quality,* 1975.

42. EPA, *Health Consequences of Sulfur Oxides.*

43. Whittenberger et al., "Review of the CHESS Program," 316–24. On the SAB, see Jasanoff, *Fifth Branch,* 84–100; Smith, *Advisers,* 68–100.

44. Whittenberger et al., "Review of the CHESS Program," 317–18.

45. Ibid., 318.

46. CHESS, *Investigative Report*, 34–39.

47. Ibid., 25–27.

48. Rood, "US Chamber Asks Sulfur Case Probe"; US Congress, *Conduct of EPA's CHESS Studies.*

49. US Congress, *Conduct of EPA's CHESS Studies*, 11–39; *Detailed Critique.*

50. Electric Power Research Institute, *Evaluation of CHESS: New York Asthma Data 1970–1971*; Roth, *Evaluation of CHESS: New York Asthma Data 1971 to 1972*; Olsen, *Evaluation of CHESS: Utah Asthma Study, 1971–1972.* On Roth's other research in industry's defense, see Michaels, *Doubt*, 50–51.

51. EPA, *Status of CHESS*; Heiderscheit and Hertz, *Assessment of the CHESS Sulfate and Nitrate Data*; Bachmann, "Will the Circle," 652–97.

52. Holland et al., "Health Effects," 525–659; Shy, "Epidemiologic Evidence," 661–71; Liu and Yu, "Regional Estimates," 26–31.

53. Rood, "EPA Study."

54. Boffey, "Sulfur Pollution," 352–54.

55. US Congress, *Conduct of EPA's CHESS Studies*, 59–82.

56. EPA, *Investigative Report*, 25–27; EPA, *Planning and Management of R&D in EPA*; Environmental Research, Development and Demonstration Authorization Act, 42 USC, 4361–70; EPA, *Addendum.*

57. Talley, "Research Mission," 2; US National Research Council, *Strengthening Science*, 28.

58. Whittenberger, "Health Effects," 129–30; "A Tale of Six Cities," *Harvard Public Health Review*, accessed Feb. 4, 2013, http://diseaseriskindex.harvard.edu/review/a_tale.shtml.

59. Anderson, *Environmental Improvement.*

60. On catalytic converters, see Lee, "Engineering the Environment," 91–126; Dunn and Johnson, this volume; and Vinsel, "Federal Regulatory Management," 194–322. On scrubbers, see Lee, "Engineering the Environment," 127–69; Taylor, "Influence of Government"; and Hart and Kallas, "Alignment and Misalignment."

61. EPA, *Interagency Energy/Environment R&D Program.*

REFERENCES

American Institute of Chemical Engineers. *Pollution and Environmental Health.* New York: American Institute of Chemical Engineers, 1961.

Anderson, Frederick R. *Environmental Improvement through Economic Incentives.* Baltimore: Johns Hopkins University, 1977.

Aspen Institute. *EPA 40th Anniversary: 10 Ways EPA Has Strengthened America.* Washington, DC: Aspen Institute, 2010.

Bachmann, John. "Will the Circle Be Unbroken: A History of the US National Ambient Air Quality Standards." *Journal of Air and Waste Management Association* 57 (2007): 652–97.

Barnard, William F. *Community Health Environmental Surveillance Studies (CHESS) Air Pollution Monitoring Handbook: Manual Methods.* Research Triangle Park, NC: EPA, 1976.

Boffey, Philip M. "Sulfur Pollution: Charges that EPA Distorted the Data Are Examined." *Science* 192, no. 4237 (Apr. 23, 1976): 352–54.

Boushey, Homer A., and John V. Fahy. "Basic Mechanisms of Asthma." In "Asthma as an Air Toxics End Point," supplement, *Environmental Health Perspectives* 103, no. S6 (Sept. 1995): 229–33.

Bregman Co. *Technical Primer on Major EPA Programs: Designed for Use by EPA's Quality Assurance Community.* EPA 650-B-70-005. Washington, DC: EPA, 1970.

Clean Air Act of 1970. 42 USC. 7401–626.

Corman, Rena. *Air Pollution Primer,* 3rd ed. New York: American Lung Association, 1978.

Dear, Peter, and Sheila Jasanoff. "Dismantling Boundaries in Science and Technology Studies." *Isis* 101, no. 4 (2010): 759–74.

A Detailed Critique of the Sulfur Emission/Sulfate Health Effects Issue. San Rafael, CA: Greenfield, Attaway & Tyler, Apr. 1975.

Electric Power Research Institute. *Evaluation of CHESS: New York Asthma Data 1970–1971.* Palo Alto, CA: EPRI, 1977.

Environmental Research, Development and Demonstration Authorization Act. 42 USC. 4361–70.

"ES&T Interview: Stanley Greenfield." *Environmental Science & Technology* 5, no. 10 (Oct. 1971): 990–92.

Finklea, John F., M. F. Cranmer, Douglas I. Hammer, L. J. McCabe, V. A. Newill, and Carl M. Shy. "Health Intelligence for Environmental Protection: A Demanding Challenge." *Proceedings of the Sixth Berkeley Symposium on Mathematical Statistics and Probability* 6 (1972): 101–109.

Greenfield, Stanley. "Our Future Is Up in the Air." *Bulletin of the American Meteorology Society* 52, no. 9 (Sept. 1971): 844–48.

———. "An Overview of Research in the Environmental Protection Agency." In *Proceedings of the Interagency Conference on the Environment,* 23–33. Livermore, CA: Lawrence Livermore Laboratory, 1973.

Greenstein, Shane M. "Lock-In and the Costs of Switching Mainframe Computer

Vendors: What Do Buyers See?" *Industrial and Corporate Change* 6, no. 2 (1997): 247–73.

Hart, David M., and Kadri Kallas. "Alignment and Misalignment of Technology Push and Regulatory Pull: Federal RD&D Support for SO$_2$ and NO$_x$ Emissions Control Technology for Coal-Fired Power Plants, 1970–2000." MIT-IPC-Energy Innovation Working Paper 10–002. Cambridge, MA: MIT Industrial Performance Center, 2010.

Heiderscheit, Leo T., and Marvin B. Hertz. *An Assessment of the CHESS Sulfate and Nitrate Data During the Period RETA Performed the Chemical Analysis.* Research Triangle Park, NC: EPA, 1977.

Heimann, Harry. *Air Pollution and Respiratory Disease.* Washington, DC: US Department of Health, Education, and Welfare, Public Health Service, 1964.

Holland, W. W., A. E. Bennett, I. R. Cameron, C. du V. Florey, S. R. Leeder, R. S. F. Schilling, A. V. Swan, and R. E. Waller. "Health Effects of Particulate Pollution: Reappraising the Evidence." *American Journal of Epidemiology* 110, no. 5 (Nov. 1979): 525–659.

Irwin, Alan, Henry Rothstein, Steven Yearley, and Elaine McCarthy. "Regulatory Science: Towards a Sociological Framework." *Futures* 29, no. 1 (1997): 17–31.

Jasanoff, Sheila. *The Fifth Branch: Science Advisers as Policymakers.* Cambridge, MA: Harvard University Press, 1994.

———. "Ordering Knowledge, Ordering Society." In *States of Knowledge: The Co-Production of Science and Social Order,* edited by Sheila Jasanoff, 13–45. New York: Routledge, 2004.

Joerges, Bernward, and Terry Shinn, eds. *Instrumentation between Science, State, and Industry.* Dordrecht: Kluwer, 2001.

Kelsey, Andrea, Gene R. Lowrimore, and Jane Smith. "The Conversion of CHESS and Other Systems." In *Proceedings Number 1 of the OR&D ADP Workshop, Bethany College, WV: October 2–4, 1974,* 282–88. Washington, DC: EPA, 1975.

Kilburn, Kaye H. "In Memoriam: John F. (Jack) Finklea; Born: August 27, 1933, Died: December 22, 2000, Consulting Editor: 1986–2000." *Archives of Environmental Health* 56, no. 4 (2001): 293.

Knelson, John H. "Evidence for the Influence of Sulfur Oxides and Particulates on Morbidity." *Bulletin of the New York Academy of Medicine* 54, no. 11 (Dec. 1978): 1137–54.

Latour, Bruno. *The Pasteurization of France.* Translated by Alan Sheridan and John Law. Cambridge, MA: Harvard University Press, 1988.

Lee, Jongmin. "Engineering the Environment: Regulatory Engineering at the U.S. Environmental Protection Agency, 1970–1980." PhD diss., Virginia Tech, 2013.

Liu, Ben-Chieh, and Eden S. H. Yu. "Regional Estimates of the Morbidity Cost of Total Suspended Particulates." *Growth and Change* 11, no. 2 (1980): 26–31.

Michaels, David. *Doubt Is Their Product: How Industry's Assault on Science Threatens Your Health.* New York: Oxford University Press, 2008.

National Industrial Pollution Control Council. *Air Pollution by Sulfur Oxides.* Washington, DC: Feb. 1971.

Nelson, William C., Victor Hasselblad, and Gene R. Lowrimore. "Statistical Aspects of a Community Health and Environmental Surveillance System." *Proceedings of the Sixth Berkeley Symposium on Mathematical Statistics and Probability* 6 (1972): 125–33.

Olsen, A. R. *Evaluation of CHESS: Utah Asthma Study, 1971–1972.* Palo Alto, CA: EPRI, 1983.

Porter, Theodore M. *Trust in Numbers: The Pursuit of Objectivity in Science and Public Life.* Princeton, NJ: Princeton University Press, 1996.

Rall, David P. "Review of the Health Effects of Sulfur Oxides." *Environmental Health Perspectives* 8 (1974): 97–121.

Riggan, Wilson B., Douglas I. Hammer, John F. Finklea, Victor Hasselblad, Charles R. Sharp, Robert M. Burton, and Carl M. Shy. "CHESS: A Community Health and Environmental Surveillance System." *Proceedings of the Sixth Berkeley Symposium on Mathematical Statistics and Probability* 6 (1972): 111–23.

Rood, W. B. "EPA Study—The Findings Got Changed: Research on Sulfur's Effects on Health Stirs Power Company Furore." *Los Angeles Times*, Feb. 29, 1976. sec. part one final, A1.

———. "US Chamber Asks Sulfur Case Probe." *Los Angeles Times*, Mar. 31, 1976, B16.

Roth, H. Daniel, John R. Viren, and Anthony V. Colucci. *Evaluation of CHESS: New York Asthma Data 1971 to 1972.* Palo Alto, CA: EPRI, 1981.

Ruckelshaus, William D. "The Conquest of the Overload." In *National Environmental Information Symposium: An Agenda for Progress Held at Cincinnati, Ohio on 24–27 September 1972, Vol. 2 Papers and Report,* 2–6. Washington, DC: EPA, May 1973.

Schooley, James F. *Responding to National Needs: The National Bureau of Standards Becomes the National Institute of Standards and Technology, 1969–1993.* Washington, DC: National Institute of Standards and Technology, 2000.

Scott, James. *Seeing Like a State: How Certain Schemes to Improve the Human Condition Have Failed.* New Haven, CT: Yale University Press, 1998.

Shaller, R. B. *System Development Plan: A Systems Approach for Acquiring Air Monitoring Networks.* Washington, DC: Mitre Corporation, 1970.

Shy, Carl M. "Epidemiologic Evidence and the United States Air Quality Standards." *American Journal of Epidemiology* 110, no. 6 (Dec. 1979): 661–71.

Smith, Bruce L. R. *The Advisers: Scientists in the Policy Process.* Washington, DC: Brookings Institution, 1992.

Statewide Air Pollution Research Center. *Community Health and Environmental Surveillance Studies California Progress Report, July 1972–April 1973.* Riverside: University of California, Riverside, 1973.

Straddling, David. *Smokestacks and Progressives: Environmentalists, Engineers, and Air Quality in America, 1881–1951.* Baltimore: Johns Hopkins University Press, 1999.

Swedlow, Brendon. "Cultural Coproduction of Four States of Knowledge." *Science, Technology, & Human Values* 37, no. 3 (May 2007): 151–79.

Talley, Wilson K. "The Research Mission." *EPA Journal* 1, no. 9 (1975): 2.

Taylor, Margaret R. "The Influence of Government Actions on Innovative Activities in the Development of Environmental Technologies to Control Sulfur Dioxide Emissions from Stationary Sources." PhD diss., Carnegie Mellon University, 2001.

US Congress. *The Conduct of EPA's "Community Health and Environmental Surveillance System" (CHESS) Studies: Joint Hearing Before the Committee on Science and Technology and the Committee on Interstate and Foreign Commerce, 94th Congress.* Washington, DC: US Government Printing Office, Apr. 1976.

US Environmental Protection Agency. *Addendum to the Health Consequences of Sulfur Oxides: A Report from CHESS, 1970–1971, May 1974.* Washington, DC: EPA, 1980.

———. *Directory of EPA, State, and Local Environmental Quality Monitoring and Assessment Activities.* Washington, DC: EPA, 1974.

———. *Environmental Protection Agency: A Progress Report, December 1970–June 1972.* Washington, DC: EPA, 1972.

———. *Environmental Protection Agency's Monitoring Programs.* Washington, DC: EPA, 1973.

———. *The Environmental Protection Agency's Research Program with Primary Emphasis on the Community Health and Environmental Surveillance System (CHESS), An Investigative Report: Report Prepared for the Committee on Science and Technology, 94th Congress.* Washington, DC: EPA, Nov. 1976.

———. *EPA at Research Triangle Park, Twenty Five Years of Environmental Protection.* Research Triangle Park, NC: EPA, 1996.

———. *Health Consequences of Sulfur Oxides: Summary and Conclusions Based upon CHESS Studies of 1970–1971.* Research Triangle Park, NC: EPA, 1973.

———. *Interagency Energy/Environment R&D Program.* Washington, DC: EPA, 1977.

———. *National Environmental Research Centers.* Washington, DC: EPA, 1973.

———. *Planning and Management of R&D in EPA.* Washington, DC: EPA, June 1978.

———. *Status of the Community Health and Environmental Surveillance System (CHESS): Report to the US House of Representatives Committee on Science and Technology.* Washington, DC: EPA, Nov. 1980.

US National Research Council. *Air Quality and Stationary Source Emission Control.* Washington, DC: National Academies Press, 1975.

———. *Strengthening Science at the US Environmental Protection Agency: Research-Management and Peer-Review Practices.* Washington, DC: National Academy Press, 2000.

Vinsel, Lee Jared. "Federal Regulatory Management of the Automobile in the United States, 1966–1988." PhD diss., Carnegie Mellon University, 2011.

Whittenberger, James L. "Health Effects of Air Pollution: Some Historical Notes." *Environmental Health Perspectives* 81 (1989): 129–30.

Whittenberger, James L., George B. Hutchinson, John W. Tukey, Peter Bloomfield, and Morton Corn. "Review of the CHESS Program: A Report of a Review Panel of the Science Advisory Board-Executive Committee, Mar. 14, 1975." In US Congress, *The Conduct of EPA's "Community Health and Environmental Surveillance System" (CHESS) Studies,* 316–24.

Williams, Dennis. *The Guardian: EPA's Formative Years, 1970–1973.* EPA 202-K-93–002. EPA History Program, Sept. 1993.

8

A HEIGHTENED CONTROVERSY
Nuclear Weapons Testing, Radioactive Tracers, and the Dynamic Stratosphere

E. Jerry Jessee

On February 2, 1951, Merril Eisenbud, an industrial hygienist at the Atomic Energy Commission's (AEC) Health and Safety Laboratory (HASL) in New York State, received an urgent phone call from his colleague Henry Blair of the University of Rochester. The Eastman-Kodak Company in up-state New York had just notified Blair that the company's film manufacturing plant had detected an abnormal rise in radiation in their air intake filters following a thunderstorm. The levels of radiation were apparently not elevated enough to warrant health concerns, but they did threaten the films being produced at the plant. Eisenbud immediately suspected the source of the radiation, even if the incident did catch him by surprise. Six days earlier, the AEC had inaugurated the recently established Nevada Test Site (NTS) with a series of nuclear weapons tests code-named Operation Ranger—the first on the North American continent since Trinity in July of 1945.

After getting off the phone with Blair, Eisenbud promptly called NTS medical director Thomas Shipman to inform him of the incident. "You're crazy, Merril," Eisenbud later recalled Shipman saying, "I was out to Ground Zero, and there's no radiation out there, and you're telling me it's up in Rochester."[1] Eisenbud next contacted AEC headquarters in Washington,

DC. To his astonishment, the AEC had no program to monitor or track off-site fallout beyond the confines of the NTS. Indeed, as the report that confirmed the feasibility of testing in the southern Nevada desert confidently asserted, "The only places to worry about are those within a radius of 150 miles" of the test site.[2] Undeterred, over the next few days Eisenbud endeavored to uncover the extent of the contamination by having HASL personnel and colleagues throughout the Midwest take local radiation measurements. With this information in hand, by the end of the month Eisenbud was able to produce a map illustrating the trajectory and pattern of the Ranger fallout as the radioactive clouds traveled eastward across the continent. The map, Eisenbud later wrote, "demonstrat[ed] that surprising amounts of fallout could occur over large areas thousands of miles from a relatively small air burst."[3] In the wake of Eisenbud's findings, the commission established a continental fallout sampling program at US Weather Bureau stations to monitor radiation activity across the nation. Consisting of a square-foot adhesive film mounted on a four-foot stand to trap fallout particles, the array of "gummed-film" collecting stations began modestly with forty-five in 1953 and expanded to 120 a few years later, some of which were located outside the United States.

The AEC's concern about offsite fallout, however, stemmed less from their apprehension about the effect of radioactive fallout on human health than from the fear that the photographic film industry might sue for future damages to their products.[4] Throughout the 1950s, the AEC consistently maintained that "heavy fallout from near-surface explosions has extended only a few miles from the point of burst. The hazard has been successfully confined to the controlled areas of the Test Site."[5] As Scott Kirsch has argued, containing public fears about radioactivity produced by the tests was dependent upon the illusion of technical control of the spatial boundaries of the NTS.[6]

Such notions of radioactive containment were deeply rooted in the AEC's managerial and technocratic attitude toward the material environment. Throughout the era of atmospheric nuclear weapons testing, the AEC continually struggled with a fundamental problem: the boundaries they erected between the spaces where the bombs were tested and the rest of the world ran headlong against the mobile nature of fallout radioactivity once deposited in the material environment.[7] An essential assumption undergirding the AEC's control of offsite fallout held that detonating the weapons under ideal conditions (that is, in the absence of rain), assured that the majority of the fallout would deposit within the confines of the test site. Yet such hori-

FIG. 8.1. Lester Machta with gummed-film stand. SOURCE: US National Archives and Records Administration, Still Picture Branch, Records of the Weather Bureau RG 27, Image 27-G-1-7A-2.

zontal boundary-making relied equally on the insinuation of control on the vertical axis. If the bulk of the heavy particles in the atomic cloud were to "fall out" within the NTS, it was assumed that the remaining radioactivity would be safely dispersed over long distances, suspended long enough to allow for radioactive decay, or simply diluted. Rendered passive and static, the atmosphere functioned solely as an agent of dispersion and dilution, limiting fallout hazards in the biosphere.

This aspect of the AEC's safety discourse proved critical for allaying fears about fallout in the wake of thermonuclear testing in the mid-1950s. Unlike the "conventional" atomic bomb, thermonuclear tests produced exponentially higher-explosive yields and were, therefore, exponentially "dirtier." Equally important, because these tests injected radioactive debris high into the stratosphere, they revealed the possibility that fallout could circulate on a global scale and potentially impact the health of every human being on the planet.

Despite the sinister implications of thermonuclear testing, the AEC maintained that the fallout dangers were minimal. The primary architect of this reassuring view was University of Chicago physical chemist Willard Libby. According to Libby, the tropopause, the isothermal layer between the troposphere and the stratosphere, constituted a nearly impenetrable atmospheric boundary layer. By preventing or, at least, moderating the downward transfer of bomb debris, the tropopause was thought to reduce fallout risks by allowing radioactive particulate matter to disperse evenly throughout the convectively stable stratosphere. The stratosphere, put another way, functioned as a natural atmospheric waste "reservoir."

Libby would not go unchallenged. Beginning in the late 1950s, he faced opposition from meteorologists who charged that his model failed to conform to material and theoretical realities. Spearheaded by US Weather Bureau scientist Lester Machta (see fig. 8.1), this group centered their objections largely on newly developed atmospheric circulation theories and firmly grounded empirical data derived, ironically, from the use of radioactive fallout as a tracer to study stratospheric air motions. The model that Machta advanced to account for this new information demonstrated that stratospherically injected radioactive debris was falling faster and less uniformly than Libby's model supposed. Although the controversy between Libby and Machta was initially played out behind closed doors, their testimony as part of congressional hearings on the fallout hazard in the late 1950s pushed the stratospheric fallout question to the forefront of public debate about the risks of nuclear weapons testing. As Machta's model of vertical transboundary movement gained favor among AEC and independent scientists, it contributed significantly to the scientific rationale for ending aboveground testing in 1963.

This chapter explores how atmospheric research reshaped how scientists and the lay public conceptualized the human health risks associated with nuclear fallout. Atmospheric and nuclear scientists investigated the complex and dynamic ways the atmosphere both modulated and magnified human exposure. This fallout research provides scientific context for how we have come to perceive global-scale pollution. Recent research has acknowledged the direct debt owed to the development, during the fallout period, of new meteorological tools such as radiotracers, global monitoring networks, and high-altitude aircraft samplers for the study of global warming.[8] Indeed, these tools, along with other technological advances such as satellite remote sensing and numerical weather prediction, helped transform meteorology into a science of the globe.[9] The atmospheric research described in this chapter helped usher in modern views of the Earth as a spatially interconnected and fragile ecosphere, by showing how ostensibly disconnected nuclear events in one part of the world held global repercussions. Through this work, a globalist perspective took hold that would animate concerns about the effects of human endeavors on the health of the ecosphere.

WILLARD LIBBY, PROJECT SUNSHINE, AND THE BOUNDED STRATOSPHERE

When the AEC began testing thermonuclear weapons in 1952, the problems associated with their fallout initially appeared to most as simply a matter of scale. Trinity-type atomic bomb tests reached yields of twenty kilotons; thermonuclear weapons released roughly 500 times more explosive energy. The yield of the first H-bomb test, Ivy-Mike, was 10.4 *megatons* (MT). Such an exponential increase in explosive yield necessarily corresponded with a concomitant rise in radioactive particle production. Yet aside from the problem of magnitude, thermonuclear testing also raised new questions about fallout risks. Whereas prior AEC efforts to control fallout had been limited to reducing human exposure to short-lived gamma radiation, the enormous amounts of alpha- and beta-emitting particles associated with the radioactive debris of the H-bomb tests coupled with their potentially global reach redirected the commission's attention toward longer-term problems. If gamma radiation posed the greatest short-term threat, what effect might the continual deposition and accumulation of long-lived radionuclides in the landscape have for human health?

Interestingly, the AEC had begun to speculate about the long-term health effects of fallout relatively early, but only as they related to nuclear war. In 1949, for example, the commission launched a small-scale secret study code-named Project Gabriel to "evaluate the radiological hazard from the fallout of debris from nuclear weapons detonated in warfare . . . the major interest [of which] may lie either in local fallout or in the superimposed long range fallout from many weapons." Put simply, the report wanted to know how many atomic bombs the USSR and the United States could exchange in a war before the radioactivity from their fallout rendered the biosphere incapable of supporting human life. The answer to that question, naturally, depended on the effect of long-lived fallout particles. In the final Gabriel report, its author reasoned that strontium-90 was the critical long-term hazard because of its long half-life (twenty-eight years) and chemical similarity to calcium, which made the radionuclide uniquely bioavailable for uptake by plants and animals.[10] If ingested by humans, strontium-90 was known to concentrate in bone and, at elevated levels, produce osteosarcoma or leukemia. Based on Gabriel, the AEC estimated that roughly 100,000 nominal Trinity-type weapons could be detonated before strontium-90 levels in soils and plants reached dangerous levels.[11]

The advent of thermonuclear weapons fundamentally altered this calculus. With such an exponential increase in destructive yield, thermonuclear weapons testing over the long term raised the specter that the world might witness doomsday even without a full-scale nuclear war: thermonuclear weapons shattered the notion of a "nominal" bomb. As a result, in the summer of 1953 the AEC convened a group of some fifty scientists and radiation experts to readdress the conclusions found in Gabriel at a top-secret meeting at Rand Corporation headquarters in Santa Monica, California. The problem that the conference attendees encountered immediately was that the severity of the hazard was clouded in uncertainty; until the conference, AEC officials within the Division of Biology and Medicine had not given priority to environmental studies investigating the movement of fallout particles in the Earth's oceans, terrestrial ecosystems, and the atmosphere. Yet because these environmental realms were critical factors linking strontium-90 to human bodies, understanding the nature of the hazard was dependent upon knowing how the radionuclide was transported through the atmosphere and biosphere.

One vital aspect of this emerging "environmental" focus centered on the movement of strontium-90 in the atmosphere and its rate of fall back to earth. However, no one knew the fate of the radioactive debris from the Ivy-Mike test. Although HASL and the weather bureau set up monitoring stations across the Pacific, they detected only a fraction of the fallout.[12] Part of the detection failure centered on logistical problems posed by finding suitable sites in the vast Pacific waters. The more pressing dilemma was the fact that the cap of Mike's mushroom cloud reached to stratospheric heights, beyond the reach of aerial monitoring. The inability of the AEC to account for Mike's debris led the Rand conference attendees to imagine two possibilities for its whereabouts: either the majority of the fallout was deposited in the ocean in areas not covered by the monitoring network, or it remained suspended in the stratosphere.[13]

The AEC and Willard Libby, particularly, pinned their hopes on the latter possibility. During the meeting, Libby introduced the idea that the stratosphere acted as a kind of natural vertical "reservoir" holding strontium-90 debris. This reservoir, Libby suggested, might potentially contain strontium-90 for upwards of a decade and "reduce greatly the probability that such materials become incorporated in the biochemistry of living organisms, including man."[14] Moreover, this protracted "residence time" of strontium-90 in the stratosphere would enable the debris to become thoroughly mixed so as to "smooth out the distribution somewhat" and limit the possibility of localized "hot spot" concentrations.[15] In effect, Libby was proposing that the tropopause, the atmospheric layer between the stable stratosphere and the turbulent troposphere, functioned as a global vertical boundary limiting the descent of stratospherically injected radioactive particles to earth.

The problem, however, was that Libby's model of stratospheric fallout was pure speculation. As the authors of the report summarizing the Rand conference admitted, even at the lower levels of the atmosphere there was "very little knowledge as to how the troposphere is cleaned of debris."[16] Even less was known about the properties of the stratosphere. The best evidence for a stratospheric reservoir rested on observations of the Krakatoa volcanic eruption in 1883, which suggested that the injection of debris in the stratosphere remained aloft for years.[17] The Krakatoa explosion, however, proved little better than anecdotal evidence. What was needed, the scientists at the Rand conference concluded, was a comprehensive investigation into the manifold environmental mechanisms that contributed to the radiological contamination of the biosphere and, subsequently, human bodies. Thus was born the AEC's top-secret Project Sunshine.

If Sunshine's goal was to produce a complete picture of the behavior of strontium-90 in a global environmental system, in Libby's hands it took on a more restrictive and technocratic tenor. As Libby's comments to his fellow Sunshine researchers in a follow-up meeting to the Rand conference a year and a half later reveal, the project was rather reductionist: "We are aiming at an equation which shows us that everything checks. We should have the storage time of strontium 90 in the stratosphere. We should have the strontium 90 content of the stratosphere . . . We must have the whole world assay[ed] for strontium 90 with the objective of finding out its effect on human life."[18] Understanding the nature of the atmosphere was, of course, an important "variable" in that global equation. Yet as Libby would learn after the next thermonuclear test in 1954, not all shared his faith in an orderly stratosphere capable of being reduced to a formulaic equation.

BEAR METEOROLOGICAL COMMITTEE AND LESTER MACHTA

Although Project Sunshine remained a secret until the spring of 1956, events following the Castle Bravo thermonuclear test at Bikini Atoll in March of 1954 forced Libby to reveal its existence. With an explosive power of over fifteen MT, Bravo far exceeded its expected yield. The test also exposed Marshall Islanders and US military servicemen to enormous amounts of fallout. The crew of a Japanese fishing vessel, the *Lucky Dragon*, received lethal doses of radioactivity. While the exposures were widely reported in the press, the AEC remained relatively tight-lipped about the nature and extent of the Bravo fallout.[19] Such reticence on the part of the AEC served to fuel suspicions among scientists and the public that the AEC was not forthcoming about fallout hazards.[20] By 1956, concerns about fallout became so explosive that presidential candidate Adlai Stevenson highlighted the issue in his bid to unseat President Eisenhower.[21]

As the AEC's credibility waned following the Bravo incident, the National Academy of Sciences (NAS) announced the organization of a set of review committees to investigate fallout hazards. Dubbed the BEAR Committees (the Committees on Biological Effects of Atomic Radiation), the undertaking was extensive, comprising not only pathological and genetic committees but also ones directly related to environmental aspects such as agriculture, oceanography and fisheries, and meteorology. In reality, however, BEAR was hardly autonomous. As Jacob Hamblin has argued, the AEC worked to populate the committees with its own scientists in order to "create the appearance of its own practices being formulated by a body other than itself."[22]

The AEC swayed the conduct and tenor of the BEAR deliberations in other more subtle ways as well. The meteorological committee, chaired by weather bureau research chief Harry Wexler with his bureau colleague Lester Machta as the rapporteur, had not even met to consider the matter put before them when Libby revealed the existence of Project Sunshine and some of its findings in the January 20, 1956, issue of *Science*. Although quite vague in how he arrived at his results, Libby nonetheless maintained that the rate of fall for stratospheric strontium-90 was roughly 10 percent per year corresponding to a storage time of ten years. This was slow enough, he reasoned, to "say unequivocally that nuclear weapons tests . . . do not constitute a health hazard to the human population insofar as radiostrontium is concerned."[23] From here on, the committee found itself in the awkward position of having to contend with the scientifically authoritative stance of the AEC, bolstered by the still quite classified Sunshine data.[24]

That was only the beginning. In April, after the committee had a chance to meet and as Machta was finishing the first draft of their report, Libby notified Wexler that he wished to meet with the committee members during the upcoming American Meteorological Society meeting in early May. The object of the meeting, Wexler wrote to the committee, was to provide Libby with a forum to discuss his recent data concerning "the stratospheric storage of radioactive debris and its influence on the problem of radiostrontium fallout."[25] During the meeting, Machta and others voiced concern over Libby's ten-year storage figure. Libby, however, managed to convince Wexler at least that the issue was moot from a biological standpoint. As Wexler later wrote the committee members, whether one put the storage time at ten or five years, the uncertainties surrounding the biological effect of radiostrontium exposure "in no way invalidate the sense of Dr. Libby's conclusions."[26] In the meantime, in what was undoubtedly an attempt to head off criticism from the likes of Machta, Libby submitted a revised draft of his conclusions with supporting evidence to the NAS for publication in the organization's flagship journal. In the paper, published the following month, Libby revealed that his storage time figures were drawn from conjectural estimates of the stratospheric content of strontium-90 with the observed levels of deposition recorded by the ground-level monitoring system. Although he granted a conservative error range of ±5 years to his ten-year residence time, he further downplayed the potential five-year figure by arguing that long residence of radiostrontium in the stratosphere had the effect of mixing the radionuclide so thoroughly that its fallout pattern would be "nearly uniform over the world."[27] The effect of his model on fallout risks was clear: the stratosphere, as a bounded vertical waste reservoir, reduced the amount of strontium-90 available for incorporation in the biosphere by holding up radioactive debris and limiting the possibility of localized concentration.

Despite Libby's inclusion of putatively empirical data to support his model, the paper had little effect on Machta. In May, while Libby's paper was under review, Machta drafted a rebuttal to Libby's NAS paper for inclusion in the revised report that, if anything, sharpened his critique of Libby by taking direct aim at both the empirical and theoretical weaknesses of

Libby's model. The ground-level monitoring data, Machta pointed out, was quite unreliable owing to the poor collecting efficiencies of the gummed-film and other monitoring instruments. Indeed, Machta suspected that the system was perhaps underestimating the amount of fallout deposition by as much as 75 percent.[28] Yet, more importantly, Machta could not countenance Libby's theoretical treatment of the stratosphere as a bounded atmospheric layer, which Machta regarded as "premature" and reaching in light of the absence of data on the mechanisms of stratospheric removal.[29] Although any consideration of stratospheric removal mechanisms was, at the time, speculation, Machta argued that these uncertainties were cause for a more conservative treatment of the issue and advocated for a "mean storage time in the stratosphere [of] roughly 5 years, with an upper bound of 10 years."[30] In the face of uncertainty, he argued, precaution should rule.

Libby and Machta corresponded in an effort to reconcile their differences, but eventually NAS president Detlev Bronk called together the chairmen of the various committees and decided "not to go into great detail concerning the weapons effects of fallout."[31] Machta noted in the cover letter to his revised draft that he was sending along to the other committee members for review, "It is in my view, in this light, that pages 24–29 inclusive [i.e., the section criticizing Libby's model] be eliminated or greatly minimized in any public report."[32] In the final report published later the next month, this is precisely what happened. In its place, the committee chose instead to highlight the uncertainties of estimating fallout rates and advocated for "a continuing program to investigate this phenomenon, including actual measurements of the radioactivity in the stratosphere."[33] Libby's NAS paper, published before their final report, became the definitive statement on strontium-90 stratospheric fallout—for the time being, at least.

1957 CONGRESSIONAL HEARINGS

The BEAR Meteorological Committee would continue to meet throughout the next couple of years and eventually produce another report in 1960. In the meantime, however, Machta decided to go public with his criticism of Libby's model in a paper delivered later in November at the Washington Academy of Sciences. Although much of the paper merely recapitulated his critique of Libby's stratospheric model that had been omitted in the meteorological committee's final report, it did contain a notable addendum. Whereas the focus of debate had centered on the storage time issue, Machta also raised serious doubt about the model's assertion of gradual deposition. According to data collected in New York City during 1955, fallout levels were highest during a period of three or four months, suggesting that stratospheric fallout may have a seasonal or perhaps even a latitudinal depositional preference. Nevertheless, Machta noted tactfully, "At the moment, there is no alternative to Libby's analysis, although one may argue for changes in details."[34]

Machta and Libby continued to hash out their disagreements in correspondence throughout the remainder of 1956 and into the following spring.[35] Privately, however, Machta was not merely trying to reconcile Libby's model to the available data but was instead developing a competing one that could account for the deposition irregularities and provide a reasonable explanation for the circulation and transference of air from the stratosphere to the troposphere. One of the resources that he drew on to provide some clue to stratospheric air motions was the recent work of British scientists Alan Brewer and Gordon Dobson. Their research tracing out the circulation of stratospheric water vapor and ozone, respectively, seemed to show a general poleward circulation of stratospheric air. As these air masses reach temperate and polar regions in the colder months, furthermore, they tend to sink into the troposphere at a break in the thin mediating layer of the tropopause, causing a seasonal uptick in ozone and water vapor at lower levels. Taken together, their work formed a compelling stratospheric circulation model for Machta and one that was compatible with the growing ground-level radiostrontium data that seemed to suggest seasonality and greater deposition in the temperate latitudes.[36] Machta was not the only one taken by the correlation. In late spring, Wexler began corresponding with Brewer specifically to discuss the fallout problem.[37] Similarly, British atomic scientists at Harwell began using the model to explain the high levels of radiostrontium recorded by their monitoring system.[38]

These developments could not have happened soon enough for Machta. Earlier in the spring of 1957, the Joint Committee on Atomic Energy (JCAE), the AEC's congressional oversight committee, announced that they were to hold a series of hearings on the fallout problem in May. There was much at stake in the hearings. With the growing distrust of the AEC and the failure of the BEAR committees to provide any firm conclusions as to the nature of the fallout hazard, the JCAE, as Robert Divine has written, was stepping in "to try to compel the scientists to come forth with a satisfactory explanation of the radiation problem."[39] Machta, in hopes of avoiding a situation similar to the NAS Meteorological Committee's failure to provide the public with an alternative to Libby's model, presented his own to the JCAE.

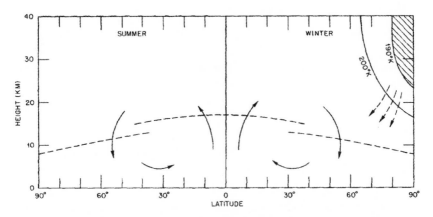

FIG. 8.2. Brewer-Dobson stratospheric circulation theory as applied by Machta. SOURCE: Machta, "Discussion of Meteorological Factors," 318.

One of the first witnesses to appear, Machta began his discussion of stratospheric fallout by offering a comparison of strontium-90 in the stratosphere to an ink blob in a bathtub:

> In mixing, one can imagine that the bathtub is stirred so that the ink quickly covers the entire water of the bathtub. In the second way, direct transport, one may imagine that a cup is dipped into the bathtub and part or all of the blob of ink is bodily lifted from one part or all of the part of the tub. It is quite evident that the first process, mixing, tends toward uniformity, whereas the second simply transports the concentration . . . I believe that the main movement of radioactive particles is the stratosphere is the result of direct transport. Mixing is so slow that the stratospheric distribution is non-uniform, even after 2 years.[40]

What's more, Machta explained, the transport of this radioactive debris circulates poleward from equatorial regions until it reaches the temperate breaks in the tropopause and begins falling rapidly in the northerly and southerly latitudes (see fig. 8.2).

These dynamic mechanisms of stratospheric circulation and tropospheric-stratospheric exchange in his model, in other words, rendered the idea of the stratosphere as a waste dump highly problematic. By way of support for his model, Machta showed the committee a graph of ground-level fallout data that demonstrated that, indeed, the majority of fallout was concentrating in the temperate latitudes of the northern hemisphere.[41]

The implication of Machta's testimony was not lost on members of the JCAE. When Libby testified, they pressed him to explain the obvious discrepancy between his and Machta's stratospheric models. Although Libby attempted to downplay their differences by declaring that the disproportionate amount of fallout was due to NTS and Soviet tests, Senator Clinton Anderson clearly perceived the repercussions that both models had for assessing fallout risks. "It affects tremendously," Anderson declared to Libby, "the question of how much fallout is safe, how much testing is safe, because if you assume that the pattern is uniform around the world, when actually it is 2 times or 3 times heavier in a given place, then you have, by this assumption, lowered the possibility of damage from fallout."[42] What was needed to settle the controversy, all agreed, were actual measurements of the movement and levels of strontium-90 in the stratosphere.

Stratospheric Monitoring and Radiotracer Experiments

The failure to sample the stratosphere for strontium-90 was not from want of trying. Shortly after the Rand conference, Merril Eisenbud developed an electrostatic precipitator that he hoped would be capable of collecting radioactive debris at high altitudes but light enough for the "sky-hook"-type balloons being developed by food manufacturing company General Mills.[43] That same year, the company was also involved with Argonne National Laboratory and Lester Machta in a balloon project at its headquarters in Minneapolis to trace the circulation of gaseous carbon-14 in the stratosphere. That project was quite successful, and indeed later analysis of the results tended to confirm Machta's stratospheric circulation model.[44] Eisenbud's precipitator failed, however, owing to its poor collection efficiency of particles in the low-atmospheric pressure of the stratosphere. The AEC terminated the flights in July of 1954.[45]

Effort to renew the project began in earnest just months after the JCAE hearings with the inauguration of the International Geophysical Year (IGY) on July 1. Remembered chiefly for its international cooperation in the quest for geophysical knowledge (buttressed, not incidentally, by the growing use of radiotracer practices in geophysical research), the IGY also served as a platform for the global expansion of the fallout monitoring network. In addition to the creation of a series of ground-level air monitors along the 80th west meridian extending from southern tip of Chile to Thule, Greenland, the IGY also included a program to develop balloon technologies to sample and measure radioactivity in the upper atmosphere both for fundamental meteorological research and to offer clues to the stratospheric content and circulation of bomb test radioactivity.[46] Out of the IGY, General Mills devised a new aerosol sampling apparatus nicknamed "ashcan," which, despite some reservations about its efficiency, prompted the AEC to establish a new balloon program titled, appropriately enough, "Project Ashcan." Soon the AEC expanded the project from its initial flights over Minneapolis and San Angelo, Texas, with two new sites at the Panama Canal zone and São Paulo, Brazil.[47]

The success of Ashcan occurred in the nick of time. In spring and summer of 1958, just before the Americans and Soviets were to enact a temporary testing moratorium, the AEC began a lengthy series of nuclear tests out in the Pacific, several of which contained two new unique radioactive tags—tungsten-185 and rhodium-102—designed to help settle the stratospheric fallout question.[48] With a half-life of seventy-four days, the tungsten tracer was injected into the atmosphere with shots conducted between April and July. The rhodium tracer (half-life 207 days) was associated with a single high-altitude test, Orange, detonated at an altitude of 26.7 miles. The tracers were added to simplify analysis of the movement of radioactive debris in the stratosphere. One of the major complexities in using strontium-90 as a tracer pivoted upon the fact that the stratospheric burden of the element was the product of several injections at various times, places, and altitudes. Although there were methods for determining the age of stratospheric debris, they were complicated and subject to high uncertainty.

Unbeknownst to the AEC, Ashcan was not the only stratospheric monitoring project currently in development in the United States. Sometime during planning for the radiotracer studies, the AEC and the NAS Meteorological Committee learned that the Department of Defense's (DoD) Armed Forces Special Weapons Project (AFSWP) had recently initiated their own stratospheric sampling program as part of the still quite secret U-2 spy plane missions. Equipped with special high-efficiency paper filters, the planes began regular north–south flights at various elevations in the stratosphere in both the northern and southern hemispheres from air bases in New York and Puerto Rico, respectively, in 1957.[49]

Dubbed the High Altitude Sampling Project, or HASP, the program had obvious advantages over Ashcan. Whereas Ashcan was limited to vertical profiles of the stratosphere at four stations, the HASP flights were able to collect fallout material along a continuous longitudinal profile at various elevations in the stratosphere (see fig. 8.3). Despite the critical interests of the commission in the flights, AFSWP kept the commission in the dark ostensibly because they would "have no security control in the AEC."[50] Mem-

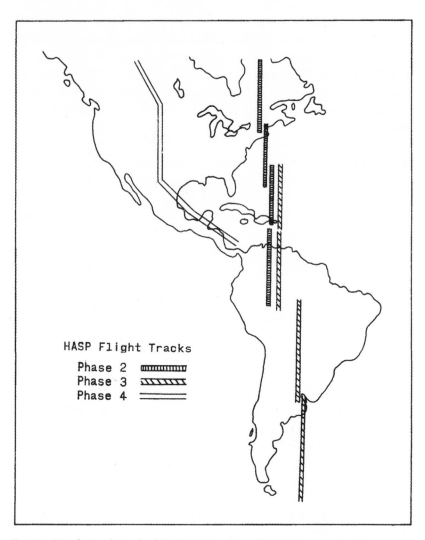

FIG. 8.3. North-South track of HASP stratospheric flights. SOURCE: Stebbins, *HASP*, 23.

bers of the NAS Meteorology Committee first learned of HASP when two AFSWP officers informed them of the project during one of their routine meetings in September of 1957. Despite its importance, however, data from HASP would not be forthcoming. Efforts by Wexler and Bronk to have the data declassified over the course of 1958 failed, apparently because AFSWP was unwilling to hand over raw, underanalyzed data.[51]

There the matter rested until December. On December 11, AFSWP for-warded the Joint Chiefs of Staff a report detailing the preliminary HASP analysis. "Recent indications," the report said, "are that the radioactivity in the stratosphere has a residence half-life of about 2 years (in contrast to the previously assumed value of about seven years)" and the "concentration of the Sr90 on the surface of the earth is greater in the United States than in any other area of the world."[52] On Christmas Eve, AFSWP officials met with AEC Commissioners and scientists to reveal the HASP findings.[53] Although Libby, not surprisingly, dismissed the import of the HASP data, he never-theless notified the JCAE of the new stratospheric data.[54]

Soon a controversy erupted among the JCAE, AEC, and DoD. On Feb-ruary 19, General Herbert Loper, assistant to the Secretary of Defense for Atomic Energy, summarized the conclusions of HASP for the JCAE on a confidential basis. In the letter, Loper informed the Committee that "tenta-tive conclusions to date indicate that three-tenths of the quantity of radioac-tive debris leaves the stratosphere each year, that the north–south diffusion of radioactive particles in the stratosphere does exist, and that in both hemi-spheres, there is a latitude band of maximum drip out which is from 35 to 50 north or south."[55] The following week, Libby wrote a rebuttal letter to Loper that he copied to the JCAE. Based on his analysis of the HASP data, Libby revised his residence time figure to four years, but insisted to Loper that it was "difficult to push it down to the 2 years you give."[56] Moreover, he argued, the "old" nonuniformity argument "still is not quite settled." He blamed local fallout from tests at the NTS for the discrepancies in hemispheric fall-out.[57] While this debate was brewing behind closed doors, Libby continued to cling to his simplified model in his public addresses. On March 13, he delivered a speech in Seattle that recapitulated the conclusions drawn in his original model. Instead of acknowledging his reinterpretation of the resi-dence time in light of new data, Libby maintained that the "rate of descent . . . is so small that something like 5 to 10 years appears to be the average time they spend before descending to the ground."[58] Upon hearing of Libby's Seattle speech, Senator Clinton Anderson, the chair of the JCAE's subcom-mittee on radiation, was infuriated. A few days later, Anderson attempted to have the content of Loper's letter declassified for a speech he was preparing to refute Libby. The DoD demurred. Although they were apparently willing to declassify the letter, Libby convinced them to withhold it owing to the preliminary nature of the data. Further enraged, Anderson subsequently ac-cused the DoD of "gagging" the JCAE when it had pertinent data to share with the American public.[59] Anderson's ploy worked; the DoD declassified the letter and on March 22 Anderson made the correspondence between the JCAE, Loper, and Libby available to the public. Anderson wrote in his press release, "it looks like strontium 90 isn't staying up in there as long as the AEC told us it would, and the fallout is greatest on the United States." Furthermore, Anderson announced that the JCAE would be holding hearings later in the spring to get to the bottom of the matter.[60]

Conducted over four days in early May, the hearings were nothing short of a disaster for Libby and his stratospheric model. The greatest damage to Libby occurred when AFSWP technical director Frank Shelton testified concerning the HASP data. Shelton reported that HASP measurements of strontium-90 since 1956, in conjunction with the detection of tungsten-185, demonstrated conclusively a preference for northern hemispheric spreading of radioactive debris. With regard to the residence time question, Shelton confirmed the estimates contained in Herbert Loper's letter to the JCAE: for shots detonated by the United States and United Kingdom in equatorial regions, the storage time for stratospheric radioactive debris was on the order of two years. Even more alarmingly, he also informed the committee that nuclear debris from Soviet thermonuclear tests conducted at Novaya Zemlya, north of the Arctic Circle, remained in the atmosphere for merely a year. Given these estimates, Shelton reasoned that if there was no further testing, the amount of strontium-90 fallout on the earth's surface would more than double by 1960, with the northern hemisphere receiving the greatest load.[61]

Following Shelton's testimony, the committee next turned to Machta to explain the atmospheric mechanisms behind the rapid rate and nonuniform pattern of stratospheric fallout. Machta reasoned, as he had in 1957, that the observed differences in fallout deposition could be explained by breaks in the tropopause in temperate latitudes as predicted by the Brewer-Dobson stratospheric circulation model. When asked what he thought of Shelton's testimony about HASP, Machta responded, "I have complete confidence in it." Moreover, he elaborated, "it is my view, [that the data] in light of Dr. Shelton's comments about the uneven distribution of tungsten 185, plus the uneven distribution at ground level air concentration, plus the uneven distribution of fallout . . . represent a fairly convincing picture . . . [that fallout] is coming out preferentially in the temperate latitudes of the Northern Hemisphere."[62]

METEOROLOGY AND THE END OF ABOVEGROUND TESTING

Perhaps because the 1959 congressional hearings were conducted during the testing moratorium, they received comparatively little media coverage.

Nonetheless, unlike 1957, what modest reporting did occur centered on the flap over the stratospheric fallout rate and its potential effects on current and future strontium-90 levels. On May 7 both the *New York Times* and *Washington Post and Times-Herald* ran articles detailing the testimony revealing the rapid descent of stratospheric fallout.[63] This current rate of fallout, when averaged, the *Times* noted, led only to a maximum concentration of strontium-90 in human bone of seven picocuries of radiation (100 picocuries of strontium-90 was considered the maximum permissible limit). But as Edward Gamarekian of the *Post* revealed after combing through data on strontium-90 levels released by the AEC during the hearings, samples of bread taken in New York City already exhibited a "radioactive strontium-90 content equal to four times the maximum 'permissible' limit."[64] That wasn't all. Later the next month, Gamarekian uncovered a series of fallout incidents throughout the midwestern United States prior to the testing moratorium that showed that strontium-90 levels in milk in the Dakotas were more than four times the world average.[65]

Gamarekian's reporting of these "hot spots" of fallout radioactivity helped to expose a fatal flaw in the AEC's assessment of the strontium-90 risk. Although research conducted by Machta and other meteorologists convincingly demonstrated that fallout was concentrated in northern latitudes, the AEC continued to average out exposures for the entire population, neglecting to account for the unusually high radioactivity levels in certain areas of the United States and therefore failing to inform the public living in these places of the extra burden of risk they were being asked to assume in the name of national security. The AEC's reliance on statistical averages, in other words, concealed vicinities of the United States where meteorological circumstances combined to intensify fallout hazards.

Throughout the next two years while the moratorium remained in place, the public would become increasingly aware of the preferential deposition of fallout in northern tier states and in particular hot spots. In large part, growing public concern over fallout concentrations stemmed from the concerted effort of independent scientists to provide the public with the most accurate and up-to-date fallout information that the AEC had in the past been unwilling or unable to supply. The leading group in what has been termed the "nuclear information movement," or "science information movement" was the grassroots organization Greater St. Louis Citizens' Committee on Nuclear Information (CNI). Founded by Washington University biologist Barry Commoner and concerned St. Louis citizens in 1958, CNI's expressed mission was the "gathering and public distribution of facts about the effects of nuclear testing and the military and peaceful uses of nuclear energy."[66] CNI was formed out of Commoner's conviction that fallout was an inherently political and moral issue, not simply a scientific one to be solved by technocrats. Shortly before forming CNI, Commoner wrote in an article in *Science* titled "The Fallout Problem," that there is "no scientific way to balance the possibility that a thousand people will die from leukemia against the political advantages of developing more efficient retaliatory weapons. This requires a moral judgment in which the scientist cannot claim a special competence which exceeds that of any other informed citizen." Therefore, the "public must be given enough information about the need for testing and the hazards of fallout to permit every citizen to decide for himself whether nuclear tests should go on or be stopped."[67] Science, for Commoner, provided the tools through which to assess hazards, but questions of policy remained a wholly public matter.

CNI, once formed, became essentially "The Fallout Problem" put into practice. Through public speaking engagements and monthly publication of current fallout figures and interpretations in its bulletin *Nuclear Information*, CNI emerged in the early 1960s as the premier clearinghouse of fallout data for the news media and members of Congress.[68] One of the primary issues that CNI focused on was strontium-90. During and following the congressional hearings in 1959, for example, CNI devoted a large portion of two of its issues of *Nuclear Information* to the stratospheric fallout controversy. Testimony delivered during the hearings, they noted, "appear[ed] to close an earlier controversy between the AEC (which maintained in 1957 that fallout would be evenly spread over the globe) and US Weather Bureau scientists who predicted localized concentration."[69] Later the next year, CNI member John Fowler commissioned Machta and his weather bureau colleague Robert List to write a chapter outlining Machta's stratospheric fallout model in a book intended for popular public consumption. In clear and accessible language, Machta and List recapitulated the data demonstrating the rapid descent of stratospheric fallout particles and their concentration in the northern hemisphere that they had presented to Congress the previous year.[70] They ended with an ominous note, however. Because of the fast rate of stratospheric fallout, they argued, short-lived fission products might pose as much a hazard to human health as slow-decaying radionuclides such as strontium-90.

Although Machta and List did not identify these potentially hazardous short-lived fission particles, they were undoubtedly referring to iodine-131.

With a half-life of only eight days, the AEC had largely ignored iodine-131, despite its known tendency to bioaccumulate in cow's milk and, when consumed, concentrate in the thyroid gland. As a result, monitoring of radioiodine in milk was instituted only after the moratorium on testing took effect. By the time the Soviets resumed testing in September of 1961, growing public fears about radioiodine soon matched strontium-90 anxieties. They had good reason to be alarmed. Between 1961 and 1963, the resumption of testing between the nuclear powers had more than doubled the amount of radioactive particulate matter in the atmosphere than had previously been accumulated in all the years prior to the moratorium in 1958.[71] Moreover, now that radioiodine was being monitored, shockingly high amounts of the radionuclide were showing up in milksheds across the country, prompting public health officials in Utah, in one well-publicized case, to order dairy farmers to switch their milk cows to uncontaminated dry feed because iodine-131 levels exceeded the maximum permissible limit.[72]

In the meantime, international pressure on the Soviets, Americans, and British from nonnuclear nations located in the higher-risk northern latitudes to end or at least regulate atmospheric nuclear testing was mounting. During the moratorium, extensive radiation monitoring systems were established in Canada, Iceland, and Scandinavia, all of which recorded high levels of fallout from Soviet tests at Novaya Zemlya. Following two Soviet tests in early November of 1962, scientists in Norway reported that "radioactive materials from atmospheric nuclear tests may be brought down to ground-level after about one week, thousands of kilometres away, in concentrations high enough to be of concern."[73] Shortly after the resumption of testing, Canada submitted a resolution to the UN General Assembly calling for further international cooperation in setting up a more geographically comprehensive fallout monitoring system and requested that the UN Scientific Committee on the Effects of Atomic Radiation (UNSCEAR) speed up its review of the fallout data for its forthcoming report.[74] As with the NAS BEAR report, UNSCEAR too had compiled a report in 1956 on the scientific evidence on the effects of fallout radiation that depended heavily on Libby's Sunshine work. The followup report issued in 1962, however, reflected and further corroborated the dynamic and nonuniform stratospheric pattern first reported by Machta. The UNSCEAR report went even further in some instances: polar shots confined to the lower stratosphere, the report contended, only stayed in the atmosphere for five to ten months.[75] With the renewal of testing, growing international consensus among scientists and the global public on the question of the short storage time, latitudinal pattern, and seasonality of stratospheric fallout had rendered Libby's model obsolete.

It was this kind of atmospheric understanding as much as the increased scale of testing in the early 1960s that animated the movement to end nuclear fallout. By 1963, the public knew substantially more about the environmental behavior of nuclear testing fallout than they did prior to the moratorium. They knew, for example, that radioactive particles produced by nuclear explosions often thousands of miles away from their homes circulated throughout the globe, rained on agricultural fields, sometimes within days, became incorporated into living things that they consumed, and settled into their bodies. And critically, they understood that they did not have a choice in the matter. The contamination of the food supply by strontium-90 and iodine-131 was simply a Cold War matter of fact; whatever one's view on nuclear testing, you could not escape fallout.

In no small measure, growing awareness of fallout levels and effects resulted from the work of scientists like Machta and grassroots organizations such as CNI to provide the American public with the data necessary to make informed decisions about the benefits and risks of nuclear weapons testing. By 1963, with two more rounds of congressional hearings (in which both Machta and CNI had participated extensively) and widespread press coverage of fallout levels, most had had enough of fallout.[76] In letter-writing campaigns and public marches and protests by newly formed anti-nuclear testing groups like the National Committee for a Sane Nuclear Policy and Women Strike for Peace, American and international politicians could no longer ignore the widespread abhorrence elicited by the inescapable fact that fallout contamination touched every region and person on the globe.[77] Although the signing of the Limited Test Ban Treaty (LTBT) in late 1963 failed to slow the arms race, by ending aboveground testing by the three great nuclear powers of the time, it was nonetheless a remarkable victory for a growing segment of the national and international populace critical of the Cold War calculus that prioritized national security concerns over human and environmental health. Prior to the testing moratorium, such a backlash against the imperative of testing was nearly unimaginable; fallout was an invisible and sensually imperceptible threat that only the most sensitive scientific instruments could detect. Yet as the world became ever more knowledgeable of the poisoning of the air, ecosystems, and bodies of humans, the risks of nuclear fallout became real and no longer seemed worth the perceived benefits.

Following the LTBT, meteorologists would continue to study the circulation of strontium-90 as well as the tungsten and rhodium tracer experiments in subsequent iterations of HASP and Ashcan.[78] These studies, however, were geared toward a more fundamental scientific goal. Fallout radiation, despite its role as a global poison, offered meteorologists an unmatched tool for studying the stratosphere. Debates about the Brewer-Dobson theory and other models of stratospheric circulation continued to swirl as data from these projects came in, but one thing remained certain: the Earth's atmosphere was a complex and dynamically integrated biosphere that resisted human efforts to simplify and control it.[79]

In an oral history conducted in the 1990s, Lester Machta was asked about his role in atmospheric nuclear weapons testing. "I think I was misled," he commented, "in underestimating what potential damage might actually have occurred from the fallout from US tests. Although by publicizing the fallout, as I did, I think the world got quite an abhorrence to nuclear testing and contributed, in my opinion, significantly to the nuclear test ban."[80] Machta understated his role. His work not only publicized an alternate view to Libby but, perhaps more importantly, signaled a profound shift in how scientists considered the ways in which the environment mediated human exposure to fallout radiation.

At the advent of nuclear weapons testing in 1945, protecting humans from the hazards of fallout was predicated, in part, by ensuring that the bombs were detonated under favorable local weather conditions. The undergirding assumption within this framework held that local weather forecasting was the key to preventing fallout outside the boundaries of the NTS. With the coming of thermonuclear testing, Willard Libby assumed, in a similar manner, that fallout could be bounded within the atmosphere, albeit on a much higher vertical plane. Yet as scientists like Machta began focusing their attention on the environment as the critical factor in assessing the risks of fallout, meteorologists turned from weather forecasting toward the fundamental study of large-scale atmospheric phenomena and mechanisms. Aided by new tools such as radiotracers, balloons, and air- and ground-level sampling systems, the notion of stratospheric containment gave way to new understandings of the atmosphere as spatially integrated and dynamic. This new way of seeing the atmosphere highlighted the idea that seemingly distant and isolated events held global significance. In this way, Machta's work with radiotracers and global data collection was part of a larger technological revolution in meteorology (e.g., numerical weather prediction, general circulation modeling, and satellite remote sensing) that was rendering the globe into a knowable object. It is no surprise, then, that the modern study of acid deposition, stratospheric ozone depletion, nuclear winter, and global climate change should follow closely on the heels of atmospheric nuclear weapons testing. Put simply, the atmosphere in the early 1960s was a profoundly different space than it was in 1945—and as a consequence, so too were our ideas about the hazards of the modern world and their effect on an increasingly fragile Earth. As the case of stratospheric nuclear fallout shows, scientific progress (and indeed technological progress in the ironic form of radiotracer tools) helped close the first great modern global environmental crisis.

NOTES

1. Eisenbud, quoted in Stannard, *Radioactivity and Health*, 963. See also Eisenbud, *Environmental Odyssey*.

2. Frederick Reines, Sept. 1, 1950, "Discussion of Radiological Hazards Associated with a Continental Test Site for Atomic Bombs," Los Alamos report no. LAMS-1173, Nuclear Testing Archive (hereafter NTA), Las Vegas, NV, accession no. NV0030434.

3. Eisenbud, *Environmental Odyssey*, 22.

4. Ibid., 67.

5. AEC, "Effects of High-Yield Nuclear Explosions," NTA, NV0049176, 7.

6. Kirsch, *Proving Grounds*.

7. On mobile nature, see Fiege, "Weedy West."

8. Edwards, "Entangled Histories"; Edwards, *Vast Machine*, 207–15.

9. Fleming, "Polar and Global Meteorology."

10. AEC Division of Biology and Medicine, July 1954, "Report on Project Gabriel," NTA, NV0720894. Quote is on page 1.

11. Hacker, *Elements of Controversy*, 182.

12. Eisenbud, *Environmental Odyssey*, 79.

13. Kramish, "Worldwide Effects of Atomic Weapons," 31.

14. Ibid., 26.

15. Ibid., 5.

16. Ibid., 49.

17. Ibid., 33.

18. Jan. 18, 1955, "Transcript of Biophysics Conference," National Archives and Records Administration(hereafter NARACP), College Park, Records of the Atomic

Energy Commission (hereafter RAEC), Record Group 326, Entry Number 73b (hereafter RG326-E73b), box 8, folder biophysics conference, 11–12.

19. Divine, *Blowing on the Wind*, 8; Eisenbud, *Environmental Odyssey*, 84–103.

20. Lapp, "Atomic Candor"; Lapp, "Fallout and Candor."

21. Divine, *Blowing on the Wind*, 72–73.

22. Hamblin, *Poison in the Well*, 18.

23. Libby, "Radioactive Fallout and Radioactive Strontium," 659.

24. Libby frequently sent his papers and public addresses to NAS president Detlev Bronk, who promptly forwarded them to the committee chairs. See, for example, Detlev Bronk to "Bill" Libby, Mar. 2, 1956, National Academies Archives (hereafter NAA), Washington, DC, Committees on Biological Effects of Atomic Radiation (hereafter CBEAR), Cooperation with the AEC, 1956–1958.

25. Harry Wexler to Members of the NAS Study Group on Meteorological Aspects of the Effects of Atomic Radiation, Apr. 19, 1956, NAA, CBEAR, Meteorological, Summary Reports, Drafts Apr. 1956.

26. Harry Wexler to Members of the NAS Study Group on Meteorological Aspects of the Effects of Atomic Radiation, May 3, 1956, NAA, CBEAR, Meteorological, folder General, 1956.

27. Libby, "Radioactive Strontium Fallout," 380.

28. Lester Machta, May 28, 1956, "Preliminary Report of the Study Group on Meteorological Aspects of the Effects of Atomic Radiation," NAA, CBEAR, Meteorology, Drafts, folder May 1956, 24.

29. Willard Libby to Lester Machta, May 18, 1956, NARACP, RAEC, RG326, Entry # UD-UP 13 (hereafter RG326-EUD-UP 13), office files of Willard P. Libby (hereafter Libby), box 1, folder reading file, Jan.–May 1956.

30. Lester Machta, May 28, 1956, "Preliminary Report of the Study Group on Meteorological Aspects of the Effects of Atomic Radiation," NAA, CBEAR, Meteorology, Drafts, folder May 1956.

31. Ibid.

32. Ibid.

33. "Biological Effects of Atomic Radiation," 60.

34. Machta, "Meteorological Factors," 177.

35. See correspondence in NARACP, RAEC, RG326-EUD-UP 13, Libby, box 1, folder reading file Dec. 1956 to June 1957.

36. Brewer, "Evidence for a World Circulation"; Dobson, "Origin and Distribution."

37. This correspondence can be found in the Harry Wexler Papers, Library of Congress, Manuscript Division, Washington, DC, box 9, general correspondence, 1957–58.

38. Stewart et al., "World-Wide Deposition."

39. Divine, *Blowing on the Wind*, 129.

40. US Congress, *Nature of Radioactive Fallout*, 148.

41. Machta's testimony can be found in US Congress, *Nature of Radioactive Fallout*, 148–61.

42. US Congress, *Nature of Radioactive Fallout*, 1217.

43. Eisenbud, *Environmental Odyssey*, 79. See also US Congress, *Fallout from Nuclear Weapons Tests*, 593. General Mills got into the business of research ballooning when they realized that their polyethylene cereal bags could be applied to balloons to improve both their height and payload capacity.

44. Hagemann et al., "Stratospheric Carbon-14."

45. US Congress, *Fallout from Nuclear Weapons Tests*, 593.

46. Lockhart and Patterson, "Measurements of the Air Concentration"; Winckler, "Balloon Study."

47. US Congress, *Fallout from Nuclear Weapons Tests*, 592.

48. J. Z. Holland, "AEC Atmospheric Radioactivity Studies," NTA, NV0137293. For a sampling of the results of the tungsten and rhodium tracers see Feely and Spar, "Tungsten-185 from Nuclear Bomb Tests"; Hoerlin, "United States High-Altitude Test Experiences"; Kalkstein, "Rhodium-102 High-Altitude Tracer Experiment"; List, Salter, and Telegadas, "Radioactive Debris as a Tracer"; List and Telegadas, "Using Radioactive Tracers"; Machta, "Transport in the Stratosphere"; Martell, "Tungsten Radioisotope Distribution"; Telegadas and List, "Global History."

49. US Congress, *Fallout from Nuclear Weapons Tests*, 774; Shelton, *Reflections*, ch. 7, 50–55. See also Stebbins, *HASP*.

50. Shelton, *Reflections*, ch. 8, 5.

51. See correspondence in NAA, CBEAR, Meteorology, General, folders 1957 and 1958.

52. Memorandum for: The Chairmen, Joint Chiefs of Staff, December 11, 1958, "Status Report on Fallout," NTA, NV0757969.

53. J. Z. Holland to Files, Dec. 24, 1958, "AFSWP Briefings on HASP," NTA, NV0025839.

54. US Congress, *Fallout from Nuclear Weapons Tests*, 2540.

55. Ibid., 2538.

56. Ibid.

57. Ibid., 2539.

58. Ibid., 2227.

59. Anderson and Ramey, "Congress and Research," 87; Gamarekian, "Defense and AEC Clash."

60. US Congress, *Fallout from Nuclear Weapons Tests*, 2536–37.

61. Ibid., 763–78.

62. Ibid., 784.

63. "Atom Test Rate"; Gamarekian, "AEC Reveals."

64. Gamarekian, "AEC Reveals."

65. Gamarekian, "Serious Fallout Cases."

66. CNI's mission can be found in any edition of its journal, *Nuclear Information*.

67. Commoner, "Fallout Problem," 1025.

68. Egan, *Barry Commoner*, 64.

69. "Concentration of Fallout," 6.

70. Machta and List, "Global Pattern of Fallout."

71. US Congress, *Fallout, Radiation Standards*, 10.

72. Hacker, *Elements of Controversy*, 221; "Radioactive Content of Milk."

73. Hvinden, Lillegraven, and Lillesaeter, "Passage of a Radioactive Cloud," 950.

74. Higuchi, "Radioactive Fallout," 381–82.

75. UNSCEAR, *1962 Report*.

76. US Congress, *Fallout, Radiation Standards*; US Congress, *Radiation Standards*; Toth, "US and Soviet Tests"; Finney, "Congress Told."

77. Katz, *Ban the Bomb*; Swerdlow, *Women Strike for Peace*.

78. J. Z. Holland, "AEC Atmospheric Radioactivity Studies," NTA, NV0137293; Feely and Spar, "Tungsten-185 from Nuclear Bomb Tests"; Hoerlin, "United States High-Altitude Test Experiences"; Kalkstein, "Rhodium-102 High-Altitude Tracer Experiment"; List, Salter, and Telegadas, "Radioactive Debris as a Tracer"; List and Telegadas, "Using Radioactive Tracers"; Machta, "Transport in the Stratosphere"; Martell, "Tungsten Radioisotope Distribution"; Telegadas and List, "Global History."

79. On atmospheric control see Fleming, *Fixing the Sky*.

80. Lester Machta, interviewed by Julius London, Oct. 31, 1993, transcript, American Meteorological Society Tape Recorded Interview Project, The National Center for Atmospheric Research Archives, Boulder, CO, 7.

REFERENCES

Anderson, Clinton P., and James T. Ramey. "Congress and Research: Experience in Atomic Research and Development." *Annals of the American Academy of Political and Social Science* 327 (1960): 85–94.

"Atom Test Rate Called Perilous." *New York Times*, May 4, 1959, 4.

"The Biological Effects of Atomic Radiation: Summary Reports." Washington, DC: National Academy of Sciences, 1956.

Brewer, A. W. "Evidence for a World Circulation Provided by the Measurements of Helium and Water Vapour Distribution in the Stratosphere." *Quarterly Journal of the Royal Meteorological Society* 75 (1949): 351–63.

Commoner, Barry. "The Fallout Problem." *Science* 127 (1958): 1023–26.

"Concentration of Fallout in the Northern Temperate Zone." *Nuclear Information* May–June 1959.

Divine, Robert A. *Blowing on the Wind: The Nuclear Test Ban Debate, 1954–1960.* New York: Oxford University Press, 1978.

Dobson, G. M. B. "Origin and Distribution of the Polyatomic Particles in the Atmosphere." *Proceedings of the Royal Society of London. Series A, Mathematical and Physical* 236 (1956): 187–93.

Edwards, Paul N. "Entangled Histories: Climate Science and Nuclear Weapons Research." *Bulletin of Atomic Scientists* 68 (2012): 28–40.

———. *A Vast Machine: Computer Models, Climate Data, and the Politics of Global Warming.* Cambridge, MA: MIT Press, 2010.

Egan, Michael. *Barry Commoner and the Science of Survival: The Remaking of American Environmentalism.* Cambridge, MA: MIT Press, 2007.

Eisenbud, Merril. *Environmental Odyssey: People, Pollution, and Politics in the Life of a Practical Scientist.* Seattle: University of Washington Press, 1990.

Feely, Herbert W., and Jerome Spar. "Tungsten-185 from Nuclear Bomb Tests as a Tracer for Stratospheric Meteorology." *Nature* 188 (1960): 1062–64.

Fiege, Mark. "The Weedy West: Mobile Nature, Boundaries, and Common Space in the Montana Landscape." *Western Historical Quarterly* 36 (2005): 22–47.

Finney, John H. "Congress Told New Soviet A-Tests Could Bring Milk Hazard." *New York Times,* June 8, 1962, 5.

Fleming, James Rodger. *Fixing the Sky: The Checkered History of Weather and Climate Control.* New York: Columbia University Press, 2010.

———. "Polar and Global Meteorology in the Life of Harry Wexler, 1933–62." In *Globalizing Polar Science: Reconsidering the International Polar and Geophysical Years,* edited by Roger D. Launius, James Rodger Fleming, and Davis H. DeVorkin, 225–41. New York: Palgrave, 2010.

Fowler, John M. "Strontium-90 in the Atmosphere and the Earth." *Nuclear Information,* Apr. 1959, 3–4.

Gamarekian, Edward. "AEC Reveals New York Bread Exceeded Strontium-90 Limit." *Washington Post and Times-Herald,* May 7, 1959, A10.

———. "Defense and AEC Clash on Fallout Rate." *Washington Post and Times-Herald,* Mar. 22, 1959, A10.

———. "Serious Fallout Cases Uncovered in Middle West." *Washington Post and Times-Herald,* June 7, 1959, A1.

Hacker, Barton C. *Elements of Controversy: The Atomic Energy Commission and Radiation Safety in Nuclear Weapons Testing, 1947–1974.* Berkeley: University of California Press, 1994.

Hagemann, French, James Gray Jr., Lester Machta, and Anthony Turkevich. "Strato-spheric Carbon-14, Carbon Dioxide, and Tritium." *Science* 130 (1959): 542–52.

Hamblin, Jacob Darwin. *Poison in the Well: Radioactive Waste in the Oceans at the Dawn of the Nuclear Age.* New Brunswick, NJ: Rutgers University Press, 2008.

Hewlett, Richard G., and Jack M. Holl. *Atoms for Peace and War, 1953–1961: Eisen-hower and the Atomic Energy Commission.* Berkeley: University of California Press, 1989.

Higuchi, Toshihiro. "Radioactive Fallout, the Politics of Risk, and the Making of a Global Environmental Crisis." PhD diss., Georgetown University, 2011.

Hoerlin, Herman. "United States High-Altitude Test Experiences, Report LA-6405." Los Alamos: Los Alamos Scientific Laboratory, 1976.

Hvinden, T., A. Lillegraven, and O. Lillesaeter. "Passage of a Radioactive Cloud over Norway, November 1962." *Nature* 202 (1964): 950–52.

Kalkstein, M. I. "Rhodium-102 High-Altitude Tracer Experiment." *Science* 137 (1962): 645–52.

Katz, Milton S. *Ban the Bomb: A History of SANE, the Committee for a Sane Nuclear Policy, 1957–1985.* New York: Greenwood Press, 1986.

Kirsch, Scott. *Proving Grounds: Project Plowshare and the Unrealized Dream of Nuclear Earthmoving.* New Brunswick, NJ: Rutgers University Press, 2005.

Kramish, Arnold. "Worldwide Effects of Atomic Weapons: Project Sunshine." RAND Corporation, 1953.

Lapp, Ralph E. "Atomic Candor." *Bulletin of Atomic Scientists* 10 (1954): 312–14.

———. "Fallout and Candor." *Bulletin of Atomic Scientists* 11 (1955): 170, 200.

Libby, W. F. "Radioactive Fallout and Radioactive Strontium." *Science* 123 (1956): 657–60.

———. "Radioactive Strontium Fallout." *Proceedings of the National Academy of Sciences of the United States of America* 42 (1956): 365–90.

List, Robert J., Leonard P. Salter, and Kosta Telegadas. "Radioactive Debris as a Tracer for Investigating Stratospheric Motions." *Tellus* 28 (1966): 345–54.

List, Robert J., and Kosta Telegadas. "Using Radioactive Tracers to Develop a Model of Circulation of Stratosphere." *Journal of the Atmospheric Sciences* 26 (1969): 1128–36.

Lockhart, L. B., and R. L. Patterson. "Measurements of the Air Concentration of Gross Fission Product Radioactivity during the IGY July 1957–December 1958." *Tellus* 12 (1960): 298–307.

Machta, Lester. "Discussion of Meteorological Factors and Fallout Distribution." In *Environmental Contamination from Nuclear Weapons Tests.* Report no. HASL-42. Washington, DC: Atomic Energy Commission, 1957, 318.

———. "Meteorological Factors Affecting Spread of Radioactivity from Nuclear Bombs." *Journal of the Washington Academy of Sciences* 47 (1957): 169–88.

———. "Transport in the Stratosphere and through the Tropopause." *Advances in Geophysics* 6 (1959): 273–88.

Machta, Lester, and Robert J. List. "The Global Pattern of Fallout." In *Fallout: A Study of Superbombs, Strontium 90, and Survival,* edited by John M. Fowler. New York: Basic Books, 1960.

Martell, E. A. "Tungsten Radioisotope Distribution and Stratospheric Transport Processes." *Journal of the Atmospheric Sciences* 25 (1968): 113–25.

"Radioactive Content of Milk Found Sharply Higher in Utah." *New York Times,* Aug. 2, 1962, 9.

Shelton, Frank H. *Reflections of a Nuclear Weaponeer.* Colorado Springs: Shelton Enterprises, 1988.

Stannard, J. Newell. *Radioactivity and Heath: A History.* Springfield, VA: US Office of Scientific and Technical Information, 1988.

Stebbins, Albert K., III., ed. *HASP: A Special Report on High Altitude Sampling Program.* Report no. DASA 532B. Washington, DC: Defense Atomic Support Agency, 1960.

Stewart, N. G., et al. "The World-Wide Deposition of Long-Lived Fission Products from Nuclear Test Explosions." In *Environmental Contamination from Weapons Tests.* New York: Health and Safety Laboratory, 1958.

Swerdlow, Amy. *Women Strike for Peace: Traditional Motherhood and Radical Politics in the 1960s.* Chicago: University of Chicago Press, 1993.

Telegadas, K., and R. J. List. "Global History of 1958 Nuclear Debris and Its Meteorological Implications." *Journal of Geophysical Research* 69 (1964): 4741–53.

Toth, Robert C. "US and Soviet Tests During '62 Doubled World's Fallout Rate." *New York Times,* June 1, 1963, 2.

United Nations Scientific Committee on the Effects of Atomic Radiation. *UNSCEAR 1962 Report.* New York, 1962.

United States Joint Congressional Committee on Atomic Energy. *Fallout from Nuclear Weapons Tests,* 86th Congress, 1st session, 1959.

———. *Fallout, Radiation Standards, and Countermeasures,* 88th Congress, 1st session, 1963.

———. *The Nature of Radioactive Fallout and Its Effects on Man,* 85th Congress, 1st Session, 1957.

———. *Radiation Standards Including Fallout,* 87th Congress, 2nd session, 1962.

Winckler, J. R. "Balloon Study of High-Altitude Radiations during the International Geophysical Year." *Journal of Geophysical Research* 65 (1960): 1331–59.

9

BURNING RAIN
THE LONG-RANGE TRANSBOUNDARY AIR POLLUTION PROJECT

Rachel Rothschild

WHEN FOSSIL FUELS are burned, certain pollutants released into the atmosphere can increase the acidity of precipitation and cause severe damage to ecosystems.[1] These pollutants can travel great distances until they are eventually deposited in rain, snow, fog, or dust. Collectively known as "acid rain," these phenomena have been observed throughout the world, but have particularly affected Scandinavia. In this chapter, I assess the development of the first international study to examine the atmospheric transport of pollutants that cause acid precipitation: the long-range transboundary air pollution project. A group of meteorologists conceived the idea for such a project in May of 1969 in light of mounting evidence that the pH of inland water bodies was increasing throughout Scandinavia. The study was conducted under the auspices of the Organisation for Economic Cooperation and Development (OECD), an intergovernmental body created in 1961 to promote economic development and cooperation among democratic, capitalist countries in Europe and North America.[2] It was formed from the Organisation for European Economic Cooperation (OEEC), which was originally created in 1948 to distribute Marshall Plan aid to Europe's war-torn countries, but also coordinated scientific research.[3]

Upon the foundation of the OECD in 1961, air pollution became a major focus of this scientific collaboration as governments struggled to improve their air quality. Through these efforts, the OECD quickly became the major international forum for member countries to work cooperatively on these problems.[4] In response to the waves of environmental activism sweeping across its member countries, the OECD created a new Directorate for Environmental Affairs on July 20, 1970, to facilitate scientific research on air pollution.[5]

Although initiated by scientists, the long-range transboundary air pollution project was funded and overseen by an emerging class of international civil servants tasked with a clear mandate to produce policy solutions to environmental problems.[6] The OECD assisted in organizing the project under the assumption that documenting the transport of pollution across national boundaries would lead to an international agreement to reduce the precursors of acid rain. However, as I will show, the negotiation of any international consensus based on the results of the OECD study was far from straightforward.

My work builds upon recent scholarship from international historians examining intergovernmental organizations formed after World War II, and historians of technology who have begun to explore the intersections of environmental history with technological and scientific developments in the twentieth century.[7] Historians such as John Krige, Amy Staples, and David Ekbladh have demonstrated that science and technology assumed increasing importance within international relations during the Cold War in areas ranging from physics to public health to the construction of dams in Vietnam.[8] Yet as historians Jeffrey Stine and Joel Tarr have noted, most of the scholarship on the intersection of environmental policy, science, and technology in the twentieth century has been produced by journalists and political scientists, with few placing these events in a historical context.[9] In his 2010 presidential address to the Society for the History of Technology, Arne Kaijser also drew attention to the need for further scholarship on these topics.[10]

As I intend to show, the long-range transboundary air pollution project can serve as a historical case study of scientific expertise in international negotiations on environmental problems. Acid rain required potentially costly reductions in pollution, which ultimately pitted major emitters against recipients of acidic deposition. This chapter begins with the origins of the OECD's involvement in air pollution research. I will examine how the OECD's experience in facilitating transnational cooperation on pollution problems during the 1950s and 1960s laid the foundation for its later engagement with the issue of acid rain. I then discuss the origins of the long-range transboundary air pollution project among Scandinavian scientists and the difficulties in executing the research. There were many constraints imposed on the study because of the need to generate policy-relevant information, as well as the Cold War rivalries between Eastern and Western Europe. Once the project began, resistance from Britain, France, and the Federal Republic of Germany, who were the major industrial polluters, threatened to inhibit accurate data collection and bankrupt the project.

Despite these setbacks, the study went forward and ultimately was effective in demonstrating the exchange of sulfur dioxide pollution between countries, particularly the transport of sulfur dioxide emissions to Scandinavia from foreign sources. With the preliminary results in hand, the environment ministers of OECD member countries gathered in the fall of 1974 to discuss issuing recommendations. Yet optimism about releasing a set of principles to guide national air pollution policies quickly diminished as the British delegates repeatedly sought to undermine the negotiations. In the final section, I review Britain's attempts to weaken a possible agreement on transboundary pollution and the subsequent turmoil it caused among the representatives, previewing the intense controversy to emerge between Norway and Britain over acid rain through the 1980s. I conclude by reflecting on the unique authority science was initially given within an economically oriented, policy-driven organization, and the consequences of this for developing international mechanisms to manage transboundary pollution, including acid rain.

DEATH-DEALING FOGS

During the first week of December 1930, a thick fog settled across the Meuse Valley in Belgium and mixed with the smoke and chemical emissions of numerous factories and power plants. Trapped by a temperature inversion, the foul airs filled the city of Liège with the noxious smell of rotten eggs. Thousands of people fell ill, and sixty-three died. A government inquiry into the cause of the "death-dealing" fog concluded that "sulphurous bodies, either in the form of sulphur dioxide or sulphuric acid" were to blame for the illnesses, and demanded the implementation of policies to prevent such accidents in the future.[11] The Liège disaster attracted attention from government officials, the scientific community, and the public throughout Europe and the United States.[12] Other similar incidents followed in American and British cities during the next two decades—St. Louis in 1939; Los Angeles in 1943; Donora, Pennsylvania, in 1948 (6,000 respiratory illnesses and 20 deaths); and the Great London Smog of 1952 (over 4,000 deaths).[13] During World War II, pollution reached such high levels in some cities that automobile headlights needed to be kept on during the daytime.[14]

Concerns about air pollution among scientists, public health officials, and national governments increased in the years following these events.[15] After the Liège disaster, scientific publications per year on atmospheric pollution doubled, and then quadrupled after the next major disaster in Donora, Pennsylvania, in 1948.[16] Several countries, notably West Germany, Britain, France, and the United States, responded to these problems by establishing new governmental bodies responsible for air pollution. In 1955, the West German Parliament established the Clean Air Commission, and in 1960, its civil code was amended to require authorization by the government for any industrial installations that might pollute the atmosphere.[17] France had already implemented regulations in 1917 that allowed the government to intervene with industrial sites posing a hazard to surrounding areas, public health, or agriculture, but in July of 1960, the president ordered the Minister of Health and Population to concentrate his efforts on air pollution.[18] The British Clean Air Act of 1956 introduced federal control over industrial emissions and mandated increased chimney heights to disperse pollutants high enough to prevent them from becoming trapped around cities during meteorological inversions.[19] In 1955, the US Congress passed legislation declaring air pollution a threat to public health and set aside funding for scientific research to investigate the problem, eventually assuming federal control over air pollution in the Clean Air Act of 1970. Environmental legislation began to sweep through member governments thereafter, with OECD member states passing thirty-one major national environmental laws between 1970 and 1975, in comparison to four laws in the period from 1956 to 1960, ten from 1960 to 1965, and eighteen from 1966 to 1970.[20]

The OEEC was one of the first intergovernmental organizations to exhibit an interest in air pollution studies, and its successor, the OECD, soon became the major international forum for its member governments to work cooperatively on the problem.[21] In light of the escalating pollution disasters, it established a committee of experts to study the best methods of measuring air pollution in January of 1957.[22] Upon its refashioning into the OECD and expansion to include the United States and Canada in 1961, air pollution became a major focus of the Committee for Scientific Research.[23] Although several other intergovernmental organizations also facilitated collaborative endeavors for a number of environmental issues, most gave minimal attention to air pollution during the 1950s and 1960s.[24]

Much of the early collaboration among OECD member states involved sharing technical expertise and scientific information about pollutants as governments struggled to improve their air quality.[25] As one example, at the request of the United States, the OECD coordinated several meetings in the late 1960s and early '70s to bring experts together from its member countries to discuss which measurement methods were being used for sulfur dioxide, the largest contributor to acid rain.[26] Norway and Sweden agreed to adopt the US method, because it seemed to meet most closely the necessary criteria of simplicity, reproducibility, sensitivity, and well-established margins of error.[27] Then, to test the technique properly, the countries of Canada, Denmark, Finland, France, West Germany, Italy, Norway, Sweden, and Britain all agreed to participate in a study organized by the United States.

The standardization of sulfur dioxide measurement was vital in laying the groundwork for the OECD's long-range transport of air pollutants project.[28] Without harmonizing their testing procedures, scientists in different countries could not be sure of the comparability of their results, a fundamental prerequisite for conducting a research study between different national groups. The OECD also created its own network of pilot observatories during the 1960s to run trials on atmospheric measurements, establishing stations with the German Research Foundation in the Federal Republic of Germany, the Research Institute of Applied Chemistry in France, the Institute for Water and Air Pollution Research in Sweden, and the Swiss Federal Laboratory for Testing of Materials in Switzerland. Because of these activities, a group of experienced laboratories was available for discussion and testing of the sampling and analysis methods during the planning phase of the long-range transport of air pollutants project, which proved crucial in preparing the appropriate methodology prior to its implementation at all participating ground stations.[29]

Through these activities, the OECD became an important forum for coordinating scientific methods and measurement among its member countries and advancing research in atmospheric chemistry throughout the late 1950s and '60s. During this same period, in response to the pollution catastrophes and "death-dealing" fogs, governments mandated the installation of higher chimney stacks in an attempt to dilute emissions and avert further incidents around cities and towns. However, industries continued to emit pollutants at ever-increasing rates, and though the policies mandating higher chimney stacks may have avoided pollution disasters, they were causing discernible effects on the environment far beyond the source of their emission.

The OECD's Acid Rain Project

In October 1967, Swedish scientist Svante Odén took the unusual step of publishing scientific findings in a popular daily newspaper before submitting them for peer review in an academic journal. Writing in *Dagens Nyheter*, Sweden's most popular daily newspaper, he argued that industrial emissions of sulfur dioxide were causing "Nederbördens försurning" (acidification of rain) and were damaging the country's ecosystems. Drawing upon over fifteen years of data from the European Air Chemistry Network (details below) and 600 lakes in Scandinavia, Odén claimed that acid rain was contributing to fish extinctions and warned of possible damage to soils and forests.[30] A huge national outcry erupted in Sweden, putting pressure on government officials to address the issue.[31] Yet a general survey of the country's factories convinced many scientists and government officials that its own fossil fuel consumption could not have caused such a marked increase in sulfur dioxide emissions. Later calculations would ultimately confirm that Scandinavia received about two-thirds of its sulfur dioxide deposition from outside sources.[32] The tall chimney stacks policies appeared to have created another environmental problem entirely by launching pollutants high into the air, allowing them to travel hundreds of kilometers downwind until they were eventually deposited in rain, snow, fog, or as dry particles.

Odén's article was responsible for generating the earliest public attention to the issue.[33] However, many Scandinavian scientists harbored concerns regarding the detrimental effects of acid deposition long before the publication of his work. As early as 1950, Norwegian biologists suggested that fish were dying due to the increasing acidity of precipitation.[34] By 1956, the leading Swedish meteorologist Carl-Gustaf Arvid Rossby noted that the pattern of high concentrations of sulfur dioxide in precipitation throughout Scandinavia suggested the possibility that British and German industrial emissions were responsible for the observed increase.[35] Rossby's results were based on data from a precipitation chemistry network created in 1948 by Swedish soil scientist Hans Egnér of the Ultuna Agricultural College in Uppsala and his assistant Erik Eriksson in order to study the input of nutrients to Scandinavian soils.[36] Rossby was so impressed by their observations that he helped expand the number of station sites to create the European Air Chemistry Network between 1952 and 1954, and recruited Eriksson to work for him in Stockholm.[37] Odén took over Eriksson's project when he left Ultuna, gathering data that further confirmed the rising acidity of freshwater lakes.[38]

The growing amount of scientific research on the increased spread of harmful gases by higher chimneys and their environmental consequences prompted national representatives to the Council of Europe's Commission for the Conservation of Nature and Natural Resources to discuss the problem.[39] The first Europe-wide scientific conference on the effects of air pollution was conceived as a result of their meeting and brought together more than one hundred scientists from fifteen countries as well as official representatives from many influential international organizations, including the OECD, World Health Organization, and the European Atomic Energy Commission.[40] In the aftermath of the Swedish uproar, the scientists gathered in April of 1968 to synthesize the growing body of ecological and meteorological data on air pollution. Participants noted that the effects of sulfur dioxide had been observed in many places far from industries and completely nonindustrialized areas were collecting rainwater with a pH as low as 2.8, about as acidic as vinegar.[41] Studies reported reduced crop yields in agriculture and forestry throughout affected areas, as well as the complete disappearance of certain species, such as lichens.[42] Based on the research presented, the scientists concluded that it was not possible to delineate limits of sulfur dioxide that would prevent injury to plants, as even minute levels were shown to be harmful.[43] Although the participants acknowledged that large gaps in scientific understanding remained, a consensus emerged that transboundary air pollution was causing such significant damages that "governments must take action."[44] The following year in May of 1969, an ad hoc meeting between meteorologists from Sweden, Norway, Britain, France, Germany, Finland, and the Netherlands at the OECD echoed these sentiments, and initiated the first attempt to try to understand the underlying atmospheric processes of acid deposition.

Eriksson, who served as Sweden's delegate to the OECD and maintained a close working relationship with Odén, presented a report at the meeting on the current evidence for increasing amounts of sulfur compounds in the atmosphere, which elicited an informal discussion of organizing a project to study the mass transport of acidic pollutants throughout Europe.[45] Brynjulf Ottar, who represented Norway at the meeting, seems to have been the most enthusiastic about the idea. That month, he was finalizing preparations and hiring of personnel for the newly formed Norwegian Institute for Air Research (NILU), which was set to begin operations in June. Ottar had earned a degree in chemistry from the University of Oslo in 1941 and joined the Norwegian Defense Research Institute during World War II, studying the dispersion of toxic warfare gases.[46] In the decades thereafter, his war experiences served as a foundation for studying local air pollution, such as emissions from metal smelters.[47] The publication of Odén's findings sparked an interest in the long-range transport of air pollution, and Eriksson's report to the OECD appears to have provided Ottar the opportunity to build the newly formed NILU into one of the preeminent atmospheric research institutes in Europe.[48]

After consulting with the Scandinavian Council for Applied Research, Ottar presented a proposal to the OECD's Environment Committee for a study on the atmospheric science behind acid rain. The majority of Western European countries initially indicated that they would participate, and Ottar worked with the secretariat of the OECD to develop a specific outline of the research.[49] In June of 1970 the OECD formed a steering committee of national representatives to finalize preparations for its implementation over the next two years, appointing Ottar as its director.[50] There were eleven participating countries: Austria, Belgium, Britain, Denmark, Finland, France, the Netherlands, Norway, West Germany, Sweden, and Switzerland.[51]

Apart from the International Geophysical Year of 1957–58, such a large, transnational study of atmospheric processes was unprecedented, and had the potential to enhance scientific knowledge about air pollution significantly.[52] The OECD's support of this work, however, was contingent on its potential applications to pollution policies. National representatives to the OECD's nascent Environment Committee, many of whom worked in the public health or environmental departments of their national governments, agreed to the project with the understanding that it would provide essential information for government authorities attempting to determine if reductions in emissions from its own industries would be enough to diminish or eliminate potential problems from acid rain or if an international approach was necessary.[53]

To accomplish these policy goals, the project set out to determine the relative importance of local and distant sources of sulfur dioxide to the air pollution over a region.[54] There were three main aspects to the program: ground measurement stations, aircraft sampling, and emission surveys. All three areas of data collection were needed to track pollutants from their source through the atmosphere to their final absorption in rainwater. The data collection and analysis were coordinated by Ottar and several of his colleagues at NILU, who asked participating countries to send them results at least once a month so that they might simultaneously adjust atmospheric models in order to improve the flight plans for aircraft sampling.[55]

Despite the groundbreaking nature of the study, it faced significant criticisms from the scientific community before it commenced due to the seeming distortion of its objectives for political reasons. In the course of implementing the acid rain program, the OECD solicited advice from numerous scientific organizations in addition to the team of atmospheric scientists they employed to manage the project at NILU. Some reservations were limited to the general scientific methodology of the study. For instance, Ulf Högström, a Swedish meteorologist at the University of Uppsala who had conducted considerable investigations into the atmospheric mechanisms of pollution transport, focused his critiques on the density of sampling stations and the level of accuracy that could be expected from national pollution source inventories.[56] But more serious criticisms about the overall purpose of the work came from meteorologist Bert Bolin and a number of fellow scientists at the University of Stockholm.[57] They were disappointed that the research did not include additional investigations into the ways in which sulfur oxides moved through ecosystems, processes that were very poorly understood. These scientists felt that studies of the ecological aspects of the problem, such as why the increasing acidity of lakes caused fish to die, deserved just as much consideration as the atmospheric issues.[58]

Indeed, during the original ad hoc meeting between Ottar, Eriksson, and other meteorologists to discuss the possibility of a joint study on acid rain, each country's expert representative concurred that research into the effects of acid rain on the environment should be included.[59] However, administrators at the OECD chose to omit ecological impacts from the study because they did not consider such concerns to be transnational, a key criterion for work taken up by the organization.[60] At the conclusion of the project, this decision may have partly hindered Scandinavian governments' attempts to negotiate for reductions in sulfur dioxide emissions. Even after the results confirmed the transport of pollutants to Scandinavia, several countries, notably Britain, cited the lack of direct evidence for severe environmental damages caused by acid rain to rationalize their refusal to reduce sulfur dioxide emissions.[61]

Scientists not affiliated with the OECD also criticized the study for not including any of the Warsaw Pact nations. The known circulation pattern of the atmosphere over Europe made it clear that transport from the German Democratic Republic, Poland, Czechoslovakia, and Hungary could be responsible for at least some of the pollution in the southeast of Scandinavia.[62]

Yet although Ottar raised this problem many times with the OECD, they refused to include any nonmember countries in the research. The OECD's stance on including Eastern European countries led Ottar and NILU to approach the United Nations' Economic Commission for Europe (ECE) in 1974 to take over the study before it was even finished, and the ECE began negotiating between the Soviet bloc and Western Europe for a monitoring program just as the OECD results were published in 1977.[63]

These impediments notwithstanding, the study faced additional obstacles due to noncooperation from Britain, France, and West Germany. Although it appears that all member governments were originally supportive of the project, these countries soon began to express hesitation about the work.[64] This was especially evident with Britain. Toward the end of planning process, the British delegate abruptly informed the other members of the steering committee that his government refused to participate in the study unless West Germany and France agreed to the project first. The other representatives appear to have been stunned at the position of the British government, which had expressed no reservations in any of the previous expert meetings or committee discussions. Several delegations registered their disappointment at Britain's unexpected resistance and found it downright "strange."[65] Eventually all three governments consented to participate, but further conflict occurred over funding just a few months before the pilot phase was supposed to launch in 1972. The French delegate told the committee that his country could not contribute financially until the next year because the deadline had passed to submit a proposal to Parliament, and Britain's representative reiterated its position that it would not participate until both West Germany and France came onboard.[66] These budgetary issues forced the Scandinavian governments to provide nearly all the funding for the pilot phase of the OECD study.[67] In fact, NILU was forced to absorb so much of the costs that it went considerably over its budget by the end of the project, at one point almost having the power shut off to its headquarters because it could not find the money to pay the electricity bill.[68]

Ambivalence about the study was also evident in the limited data submitted by Britain, France and West Germany. As NILU began preparations for the full yearlong measurement phase, it reported that more observatories were needed to account for areas of high rainfall in France, which had six online, but that France would not provide them with the additional stations. West Germany submitted almost no data during the trial, and the minimal data it did transmit was considerably delayed and not in accordance with the agreed-upon measurement techniques. When NILU informed the

West German delegate that ten more observatories were required to account for topographic factors in its region, the country supplied them but only equipped the stations with the absolute minimum measurement technology. Britain furnished only two stations for the trial period, and offered one more for the upcoming full measurement period.[69]

In contrast, Finland, Denmark, Sweden, and Norway all had significantly larger networks for taking observations, especially when compared to their landmass and population size. Denmark even offered to build an additional advanced station on the Faroe Islands because of their unique geographic position between Britain and Norway. They ranged from eight in Finland to almost thirty in Norway. These countries also equipped several stations with additional technology to account for the influence of other pollutants, such as nitrogen oxides, on the acidity of precipitation and employed round-the-clock sampling devices. The rest of the sampling stations in the study, including virtually all of those in Britain, France, and Germany, took measurements every six hours.[70] One station in France was equipped with such technology, but none in Britain or West Germany. The aircraft sampling during the full measurement phase further reveals the disparity among the participating countries: whereas Sweden and Norway flew thirty-five and thirty-seven flights, respectively, Britain participated in twenty-two, Germany in eight, and France in five.

Despite these asymmetries, the study was successful in generating sound estimates of each country's contribution to the pollution levels of its neighbors.[71] Five of the eleven participating nations were clearly subjected to greater amounts of air pollution from foreign countries than that produced by their own industries, with Scandinavia as a whole receiving two-thirds of its pollution from outside sources.[72] The study also made notable advances in determining the varying influences of other pollutants, such as particulate matter and nitrogen oxides, in causing acidity of precipitation. While it fulfilled its objective and found evidence that the long-range transport of sulfur dioxides occurred along with sound estimates of each country's contribution, the attempt to formulate policy recommendations based on the study's findings faltered amid heavy opposition from Britain.

THE END OF THE "HEROIC" PHASE

While the Environment Committee was formed separately from the study on long-range transport of air pollutants, the policy aspects of transboundary air pollution regulation preoccupied the delegates to its early meetings. As a result of the substantial interest shown by member governments, shortly after its inception in 1970 the committee decided to commission economic reports and legal advice from its member countries, as well as preliminary analyses by a subcommittee of economic experts.[73] These specialists, who were drawn from both the OECD staff and academic institutions throughout member countries, took the lead in developing and promoting the "polluter pays principle" as a means of assigning the costs of pollution control. The principle served to avoid distorting international trade and creating unfair competitive advantages for the industry of one country over another.[74]

Based on the fruitfulness of the investigations by the subcommittee of economic experts, in September of 1973 the Environment Committee enlarged its program of activities on transboundary pollution by setting up an ad hoc group to examine the administrative, legal, and institutional aspects of transboundary pollution.[75] As the legal and policy work progressed, a review of the results to date in the long-range transport of air pollutants study convinced the project's steering committee that the transport of sulfur dioxide was extensive and contributing to the acidification of precipitation.[76] Though the study continued for another year and a half, these results prompted the delegates to the ad hoc group to organize a high-level meeting of each country's minister of the environment for November of 1974. Similar sessions in areas such as trade and payments, energy, and agriculture had already been held.[77] The organizers of the Environment Committee hoped the meeting would move beyond the broad pledges made at the 1972 United Nations Conference on the Human Environment in Stockholm, the first major international conference on environmental issues, and establish concrete directives on environmental policies for the next decade.[78] When the ministers finally arrived in Paris for the two-day event, many of them voiced their desire to give priority to discussions of progress on transboundary pollution policies. In particular, Norwegian Minister Gro Harlem Brundtland opened the conference by arguing that, on the basis of the facts now available from the long-range transport of air pollutants project, member countries should agree to reduce their sulfur dioxide emissions. Although acknowledging the costs to such an undertaking, Brundtland believed that they would not be prohibitive in light of the development of new and cheaper pollution control technologies. Without action, she stressed, the assimilative capacity of Norwegian soil would soon near complete exhaustion.[79]

The national representatives to the OECD's Environment Committee had already spent much of the previous year drafting recommendations on ten topics for the ministers to review and finalize, ideally with minimal changes. But transboundary air pollution had proven to be an extremely contentious area, and the final recommendations were highly inadequate to fulfill the hopes of the Norwegian and other delegations. In fact, they were almost completely discarded before the meeting because of controversies among the delegates to the Environment Committee.

The committee members had finished a draft of the transboundary pollution recommendations in June of 1974, five months before the meeting was scheduled, along with the nine other declarations related to other environmental issues. However, throughout preparatory sessions in the summer and into the fall, opposition from the British delegation derailed its finalization multiple times.[80] Meeting notes and drafts of the transboundary pollution recommendations confirm that Britain repeatedly tried to alter its contents, and the document was referred back and forth between the Environment Committee and Executive Committee of the OECD several times.[81] During these deliberations, the British delegation voiced its displeasure over what it perceived as an absence of financial considerations.[82] They continually demanded the deletion of all references to pollution "beyond national frontiers," calling such statements "superfluous."[83] The chair of the Environment Committee, American physicist Hilliard Roderick, chastised the British delegation for these remarks, noting that the OECD was specifically designed to deal with problems that were international in character.[84] Roderick was a former scientist of the US Atomic Energy Commission and deputy director of the UN Natural Sciences Department during the late 1950s and '60s. He later became an outspoken critic of US policies on acid rain upon his departure from the OECD in 1978.[85]

The debates centered on the principle of "equal right of access," the fourth title of the transboundary pollution declaration. It would have given foreign parties affected by pollution the same rights to remediation for environmental damage as citizens empowered within the courts of the polluting nation.[86] This principle was proposed in conjunction with the principle of "nondiscrimination," whereby a country would agree to regulate emitters who contributed to transboundary air pollution no less stringently than domestic polluters, as well as include possible foreign environmental damage when assessing whether the total ecological impact from new polluting facilities would be too great. Many delegations hoped that the codification of these two "principles" in the legal system of each member country would be a first step in improving upon the UN Stockholm Declaration of 1972.[87] It was the first international agreement to recognize environmental harm as a human rights issue, stipulating that the protection of the environment was essential to "the enjoyment of basic human rights."[88] However, there were no binding mechanisms to ensure compliance with the principles laid out in the Stockholm Declaration.[89] The adoption of the OECD recommendations by its member countries would promote the development of the protection of the environment into "a sort of new human right" with legal enforceability, potentially liberating environmental policies from "territoriality" and state sovereignty.[90] During the preparatory discussions for the ministerial meeting, the Norwegian delegates in particular underscored the need for the same legal rights to protection and compensation for foreigners as citizens were given within the polluting country.[91]

However, the British delegates had succeeded in weakening the recommendations to such a great extent that the plan to make any declaration was almost thwarted, not by the Norwegian delegates but by the Canadian representatives, who felt that the document would turn "the clock back 65 years for the North American members of the OECD."[92] Although acknowledging that less progress had been made in Europe than between Canada and the United States, they insisted the transboundary pollution recommendations should be more than a sanctioning of existing present practices, which many delegates felt were inadequate. Lamenting the tight deadline of the ministerial meeting and the serious deficiencies of the proposal, a Canadian representative argued that the circumstances required completely jettisoning the recommendations, "whose inadequacies we may later come to regret."[93] Other delegates, notably the Americans, agreed with the Canadian assessment of the document. However, West Germany, the Netherlands, and Norway felt that the document was the "cornerstone" of the submissions to the ministers, and that removing it entirely would deprive the meeting of much of its usefulness.[94] Canada relented in the face of opposition from its colleagues who wished to include at least some declaration related to transboundary pollution, however flawed, but the delegation attached a scathing dissent as an appendix. Characterizing the statement as a "retreat from the Stockholm declaration which we all accepted two and a half years ago," the Canadian representatives asserted that the text of the document should have been preserved in its original form, particularly in reference to the section on equal right of access.[95] The controversy became so heated in the days leading up to the meeting that the document had to be transmitted to the ministers with disputed sections left in brackets and a special note attached explaining that major points of disagreement had not been resolved among the delegates to the Environment Committee.[96]

On the one hand, upon examining the UN Stockholm Declaration as compared to the OECD recommendations, the Canadian delegates were slightly overstating their case in saying that the latter did not move beyond the former in any way. The OECD principles on transboundary pollution contained more details regarding appropriate notification and consultation between countries when polluting facilities were constructed along borders, the creation of warning systems for sudden increases in emissions, exchange of monitoring measurements, and the application of the polluter pays principle to all facilities regardless of whether their emissions affected domestic or foreign peoples. It even retained a section about the issue of an equal right of access, despite the acrimonious deliberations.[97]

Yet the Canadians' frustrations were most certainly tied to the extremely watered-down language when comparing the two drafts of the OECD recommendations.[98] The original emphasized the importance of harmonizing environmental policies between member states; the final text contained no statements to that effect.[99] While the initial document stipulated that neighboring countries should have the right to request information on new pollution sources if it was not given freely, the final version eliminated those provisions.[100] Wording about the "rights" of victims was replaced with that of "no less favorable treatment." References to an "equal right" to be heard and appear in court were replaced with language encouraging countries to afford victims the option to petition for "standing in judicial or administrative proceedings."[101] This alteration left courts the option to deny hearing such cases using the principle of territoriality of laws, which some countries had interpreted to mean that foreign persons had no rights in their legal systems on the grounds that domestic legislation only applied to actions within the territory of a state and all persons or interests therein.[102] The famous Trail Smelter case of 1935, in which American farmers petitioned the US government to take action against Canada over fumes emitted from a smelting plant, is the most prominent example; Canadian courts refused to invoke jurisdiction over damage to land abroad and the case had to be settled through arbitration.[103] The OECD principles would have allowed similar future incidents to be settled in the courts of the offending nation. Regarding the principle of nondiscrimination, while the initial document called for countries to "adopt" the principle, the final version asked countries to "initially base their actions" on this principle.[104] But perhaps the most glaring change was the removal of all references to the need to reduce or eliminate transboundary air pollution and the emission of sulfur dioxides.[105]

The Norwegians were evidently unsatisfied with the end result of the OECD negotiations, and they soon took steps to draw international attention to problems with the adopted principles and press for further action. A year later Brundtland reported the results of the meeting to the Council of Europe, and though acknowledging the success of adopting a transboundary pollution declaration "not without some difficulties," she issued an appeal to take further action on the problem of sulfur dioxide emissions.[106] Brundtland argued that it had now been established "beyond all reasonable doubt" that the long-range transmission of air pollutants was occurring and contributing to serious pollution problems in a number of countries.[107] The potential damage to Norwegian lakes and rivers provided ample justification for the country's alarm over the problem, but they were also concerned about the possible long-term impact on forests and human health, she explained. The Norwegian government believed that the only solution would be to further harmonize environmental policies and develop a legally binding international instrument on acid rain, which it would continue to pursue.[108]

Despite Brundtland's exhortations to the Council of Europe, an agreement to reduce sulfur dioxide emissions was not forthcoming. In fact, just two years after the OECD's ministerial meeting, the Council of Europe passed a resolution chastising the OECD's work on transboundary pollution. It stated that "the adoption of recommendations is not enough to solve these complex problems" and that a lasting solution could only be achieved through the implementation of international legal instruments alongside scientific research and the application of any future findings to policymaking. The Council of Europe urged the OECD to cooperate with them in pursuit of these goals.[109] While the Convention on Long-Range Transboundary Air Pollution was finally ratified through the United Nations' Economic Commission for Europe in 1979, its major achievement, thanks to the tireless efforts of Ottar, was the establishment of the European Monitoring and Evaluation Program (EMEP), the cooperative program for monitoring and evaluation of the long-range transmission of air pollutants in Europe that included the Eastern European countries. The convention included no provision for reducing sulfur dioxide emissions, and the early 1980s witnessed a severe deterioration in the relationship between Norway and Britain.[110]

After the ascension of the British Conservative Party and the election of Margaret Thatcher as prime minister, research funding on sulfur dioxide and acid rain was slashed in spite of protests from scientists at home and abroad that her actions violated the stipulation in the convention of 1979 to increase financing for scientific work on the problem.[111] In response, the European Commission withdrew a number of research grants to Britain, citing the government's elimination of funding for studies on the effects of acid rain.[112] At the same time, the Thatcher administration claimed that the scientific evidence for acid rain was still uncertain, previewing tactics to stall action on other environmental threats like global warming.[113] Indeed, on several occasions, the government's Central Electricity Generating Board (CEGB) released findings that purportedly showed that acid rain caused minimal damage to Norway's ecosystems. By 1985, these actions by the CEGB resulted in tense meetings between the Norwegian and British ministers of the environment on the controversy, particularly after British scientists at the Ministry of the Environment publicly sided with Norwegian scientists against the CEGB in 1985, putting further pressure on the government to agree to pollution reductions.[114]

Norwegian scientists who had spent the previous decade engaged in research on acid rain reacted to the CEGB's publications with deep skepticism, even outrage.[115] Ottar, who had recently begun serving as the director of the chemical coordinating center for EMEP, did not mince words when discussing the controversy among his colleagues.[116] In a workshop on acid rain held in January 1983, Ottar unequivocally condemned the CEGB's calls for "more research" as a tactic to postpone emissions limits "as long as possible." For Ottar, Norwegian scientists had a responsibility to refute the CEGB in order to secure emissions reductions and prevent even more environmental degradation from acid rain.[117] The Norwegian political establishment was similarly angered over Britain's attempt to denigrate the weight of the current scientific evidence for emissions reductions, as well as reports that the government would not release results of its own studies showing that British emissions were harming the Scandinavian environment. A Norwegian government spokesman was quoted as stating that it would be "unhappy if the study were to delay measures which we now know are needed to reduce emissions."[118] Britain refused to participate in the Helsinki Protocol of 1985 to the original convention, the first international agreement to call for specific reductions in emissions of an air pollutant, which stipulated that countries reduce their sulfur dioxide emissions from their 1980 levels by 30 percent before 1993.[119] It was only in 1988 that the British government eventually signed the Sofia Protocol of 1988, which added reductions of nitrogen oxides to the 1985 agreement to reduce emissions of sulfur dioxide.[120]

In closing the 1974 ministerial meeting, Roderick remarked that the deliberations reflected the onset of a new stage in environmental policies quite different from the earlier "enthusiasm." The first "heroic phase," which had seen the establishment of environmental authorities in different countries, had ended. Difficult policy choices were now necessary, with the potential for conflicts to emerge between industries, economic growth, scientific research, and environmental impacts.[121] The meeting's shortcomings prompted the OECD to reexamine the intellectual underpinnings of the notion of "environmental rights," and a few months after the November gathering, the OECD formed the Transfrontier Pollution Group as a subdivision of the Environment Committee designed to "intensify cooperation and stimulate action" on transboundary pollution.[122]

The history of the OECD's long-range transboundary air pollution project thus reveals both the authority science had in the early years of international collaboration on environmental policies and its many limitations. Ottar was given the opportunity to organize a study of unprecedented scale on acid rain, but had little recourse to set the agenda, police the countries withholding financial contributions, or rectify their poor participation. The findings of the OECD study were never in dispute during the 1974 debates over the transboundary air pollution recommendations. Difficulties in reaching consensus on how to best cooperate internationally to address the problem, the appropriate legal protections for victims of transboundary pollution, or whether such collaboration should even occur proscribed the possibility of reducing sulfur dioxide. Under the assumption that the "facts" would lead the day, policymakers and delegates to the OECD were caught unprepared for the degree of subterfuge and resistance from Europe's heavy polluters. In short order the "facts" too would become objects of contestation, but for this moment in the early 1970s, European environmental policymakers and diplomats placed their hopes in atmospheric scientists' abilities to measure and sample the skies. As a result, knowledge about the transport of air pollution was substantially increased—though as Ottar lamented, in the end the research was not enough to prompt international action on the environmental threat of acid rain.

NOTES

Archival research for this chapter was supported by a National Science Foundation Travel Research Grant. I would like to thank the graduate students and faculty in Yale's Program in the History of Science and Medicine for their helpful comments on an earlier draft, with special thanks to Daniel Kevles, Bruno Strasser, Helen Curry, Robin Scheffler, and Naomi Rogers. I am also greatly indebted to Patrick Cohrs at Yale for many stimulating conversations on approaches to this topic. Finally, I am grateful to all the participants at the Gordon Cain Conference for their feedback and suggestions, particularly Licia Peck, Ann Johnson, and Jim Fleming for their advice on revisions.

1. Hesthagen, Sevaldrud, and Berger, "Assessment of Damage."

2. Long, *International Environmental Issues.*

3. OECD, *International Scientific Cooperation.*

4. OECD, *Methods of Measuring*; US Congress, *1972 Survey.*

5. "Environment Committee: 1st Session," Nov. 4, 1970, Organization for Economic Co-operation and Development Archives, Paris, France (hereafter cited as OECD Archives), ENV/A(70)3 Annex I, 7. See also OECD, *OECD and the Environment*, 7.

6. Staples, *Birth of Development.*

7. Stine and Tarr, "Intersection of Histories," 601–40.

8. Krige, *American Hegemony*; Staples, *Birth of Development*; Ekbladh, *Great American Mission.*

9. Stine and Tarr, "Intersection of Histories," 636–37.

10. Kaijser, "Trail from Trail," 137.

11. "Atmospheric Pollution," 277; Haldane, "Atmospheric Pollution and Fogs," 366–67.

12. Keyes, "Pure Air," 427–30; Talman, "Death-Dealing Fogs"; Firket, "Sur les causes"; Allix, "A propos des brouillards"; "Air Pollution," 1312–13.

13. Bell, Davis, and Fletcher, "Retrospective Assessment," 6–8.

14. Ahrens, *Meteorology Today*, 490.

15. Dalton, *Green Rainbow*, 33–36; Dewey, *Don't Breathe the Air.*

16. Halliday, "Historical Review," 11.

17. OECD, *Air Pollution*, 77–78.

18. Ibid., 81–82.

19. Ibid., 86.

20. Long, *International Environmental Issues*, 13; US Congress, Senate, Committee on Commerce, *1972 Survey*, 104.

21. OECD, *International Scientific Cooperation*, 62; US Congress, Senate, Committee on Foreign Relations, *Background Documents*, 6.

22. OECD, *Methods of Measuring Air Pollution.*

23. Ibid., 7–10.

24. US Congress, Senate, Committee on Commerce, *1972 Survey*, 75–99.

25. Ibid.

26. "Group of Air Pollution Experts on Measurement of Air Pollution," Oct. 25, 1971. OECD Archives, NR/ENV/71.45, 2.

27. Ibid., 6.

28. Ibid., 5.

29. Ottar, "Assessment of the OECD Study," 447.

30. OECD, *Ad Hoc Meeting*; Odén, "Acidification of Air"; and Odén, "Nederbördens försurning"; cited in Cowling, "Acid Precipitation," 110–23A.

31. Bolin, *Air Pollution.*

32. Ottar, "Long Range Transport of Sulfurous Aerosol," 513.

33. Sivertsen, *NILU; 40 år i lufta,* 27–28.

34. Ibid.

35. Rossby, "Current Problems," 40–41.

36. Rodhe, "Human Impact."

37. Bolin, "Carl-Gustaf Rossby."

38. Hessam Taba, interview with Erik Eriksson, *WMO Bulletin* 47, no. 4 (Oct. 1998), 327, http://www.wmo.int/pages/publications/bulletin_en/interviews/eriksson_en.html.

39. Harcourt, "Intervention," 9.

40. ten Houten, *Air Pollution,* 5, 7, 407–11.

41. Ibid., 137–41, 179–80.

42. Ibid.

43. Ibid., 379.

44. Ibid., 7.

45. OECD, *Ad Hoc Meeting*; OECD, *Air Management Problems,* 145.

46. Øystein Hov, interview with the author, May 19, 2011.

47. Ibid.

48. Sivertsen, *NILU; 40 år i lufta.*

49. OECD, *Air Management Problems,* 145, 175–90.

50. Ibid.

51. Ibid.

52. On the International Geophysical Year and its predecessors see Launius, Fleming, and DeVorkin, *Globalizing Polar Science.*

53. "Environment Committee: Proposal for a Co-Operative Technical Project to Measure the Long Range Transport of Air Pollutants," Sept. 7, 1971, OECD Archives, ENV(71)28. 82.075, 3.

54. "Co-Operative Technical Programme to Measure the Long-Range Transport of Air Pollutants: Note by the Secretary General," Feb. 24, 1972. OECD Archives, /M(72)7(Prov.) Part I, 4.

55. "Preliminary Outline of the Technical Work in Relation to a Proposed Project to Study the Long Range Transport of Air Pollutants," Oct. 25, 1971, OECD Archives, NR/JAH/A.71.163, Annex to Letter, 2–3

56. Ulf Högström, "Short Comments on a Proposed OECD-Project to Study

the Regional Acidification of Rain-Water," OECD Archives, NR/ENV "DIVERS," E.44.672.

57. Bert Bolin, Lennart Granat, and Henning Rodhe, "Comments to 'Outline Plan for a Project Study: Long Range Transport of Air Pollutants," OECD Jan 1971, OECD Archives, NR/ENV "DIVERS," E.44.672.

58. Ibid.

59. OECD, *Ad Hoc Meeting*, 3.

60. "Air Management Sector Group: Air Management in Relation to Emissions of Sulphur Oxides. Committee Room Document No. 1," Oct. 25, 1971, OECD Archives, E.48.414, 2.

61. "Transfrontier Pollution Group: Summary Record of the 13th Meeting Held on 23–25th October, 1978," Nov. 29, 1978, OECD Archives, ENV/TFP/M/78.13, 2.

62. Ibid.

63. Ottar, "Organization of Long Range Transport," 220.

64. Ibid., 221–22.

65. "Executive Committee: Summary Record of the 306th Meeting" (n. 66), 7.

66. Ibid., 8–9.

67. Ibid.

68. NILU, "NILU gjennom 25 år," 15–16, Norwegian Institute for Air Research, Archives and Library (hereafter cited as NILU Archives and Library), Oslo, Norway.

69. "Cooperative Technical Programme to Measure the Long-Range Transport of Air Pollutants. Steering Committee: Summary Record of the Third Meeting," Dec. 14, 1973, OECD Archives, NR/ENV/73.62, 12–19.

70. Ibid.

71. Ottar, "Monitoring Long-Range Transport."

72. Ibid.; *OECD Programme on Long Range Transport*, chapters 9, 11.

73. "Environment Committee: Summary Record of the 3rd Session," Nov. 2, 1971, OECD Archives, ENV/M(71)2, 6.

74. "Environment Committee: Issues Identified by the Sub-Committee of Economic Experts for Consideration by the Committee," Sept. 21, 1971, OECD Archives, ENV(71)30.

75. "Environment Committee: Summary Record of the 9th Session," Sept. 24, 1973, OECD Archives, ENV/M(73)3, 8.

76. *OECD Programme on Long Range Transport*, 1–6.

77. OECD, *International Scientific Cooperation*, 64.

78. "Environment Committee: Meeting at the Ministerial Level: Minutes of the 13th Session," Mar. 6 1975, OECD Archives, ENV/M(74)4. 12.484, 6–8. United Nations, *Environmental Conventions*.

79. "Environment Committee: Meeting at the Ministerial Level. Minutes of the

13th Session Held at the OECD Headquarters in Paris on the 13th and 14th November 1974," Mar. 6, 1975, OECD Archives, ENV/M(74)4 & Corrigendum, 23.

80. "Environmental Committee: Draft Action Proposals for Submission to the Environment Committee at Ministerial Level," June 19, 1984, OECD Archives, Annex VIII to ENV/MIN(74)7. 6.882, 2. Also see "Preparation of the Environment Committee Meeting at Ministerial Level," July 1 1974, OECD Archives, ENV/M(74)2, 7.

81. "Air Management Sector Group: Note by the Secretariat," Mar. 12, 1974, OECD Archives, Addendum 1 to NR/ENV/74.9. E.64293, 3.

82. "Council: Draft Declaration on Environmental Policy Submitted to the Environment Committee at Ministerial Level," Sept. 23, 1974, OECD Archives, C/M(74)21 (Prov.), 12–13.

83. "Executive Committee: Summary Record of the 380th Meeting," Oct. 3, 1974, OECD Archives, CE/M(74)16 (Prov.), 4.

84. Ibid., 4–5.

85. Roderick, "Future Natural Sciences Programme"; Schmandt and Roderick, *Acid Rain*.

86. "Executive Committee: Summary Record of the 381st Meeting," Oct. 1, 1974, OECD Archives, CE/M(74)17(Prov.) Part I, 3.

87. "Council: Action Proposal on Principles Concerning Transfrontier Pollution /C(74)158 Annex X (1st Revision) dated 4th October, 1974/ Statement by the Delegate for Canada at the Meeting of the Council on 8th October, 1974," Oct. 9, 1974, OECD Archives, CES/74.95, 1–4.

88. United Nations, "Declaration of the United Nations Conference on the Human Environment, Stockholm, June 16, 1972," UN Doc. A/CONF.48/14/Rev.1 (1972).

89. Sohn, "Stockholm Declaration," 427.

90. Environment Committee Transfrontier Pollution Group, "Mandate of the Group," Sept. 9, 1976, OECD Archives, ENV/TFP/76.18, 29.

91. "Annex VI: Proposals for a Draft Declaration by the Delegation for Norway," Apr. 16, 1974, OECD Archives, ENV/MIN(74)6, 19.

92. "Environment Committee: Meeting at the Ministerial Level" (n. 80), 46.

93. Ibid.

94. Ibid., 7–9.

95. "Council: Action Proposal" (n. 88).

96. "Procedure to Be Followed for the Adoption by the Council of Certain Texts Approved by the Environment Committee at Ministerial Level," Oct. 24, 1974, OECD Archives, C/M(74)24(Prov.) Part I, 14–15; "Environment Committee: Sum-

mary Record of the 12th Session," Sept. 11–12, 1974, OECD Archives, ENV/M(74)3 & Corrigendum, 5–6.

97. OECD, *Legal Aspects of Transfrontier Pollution*, 13–18.

98. "Environment Committee: Draft Action Proposals for Submission to the Environment Committee at Ministerial Level," May 8, 1974, OECD Archives, Annex VIII to ENV/MIN/74.7, 1–9.

99. Ibid., 4.

100. Ibid., 5.

101. Ibid., 7.

102. OECD, *Legal Aspects of Transfrontier Pollution*, 125.

103. Environment Committee Transfrontier Pollution Group, "The Role of Domestic Procedures in Transnational Environmental Disputes, by Peter Sand, United Nations FAO Legal Office," May 29, 1976, OECD Archives, ENV/TFP/ "DIVERS," 14.

104. Environment Committee: Draft Action Proposals for Submission to the Environment Committee at Ministerial Level," May 8, 1974, OECD Archives, Annex VIII to ENV/MIN/74.7, 6, Published in OECD, *Legal Aspects of Transfrontier Pollution*, 7–16.

105. Ibid., 5. "Council: Minutes of the 368th Meeting," Oct. 17, 1974, OECD Archives, C/M(74)22(Prov.), 6.

106. "Environment Committee: Statement by Mrs. Gro Harlem Brundtland, Minister for the Environment in the Norwegian Government, at the 26th Session of the Parliamentary Assembly of the Council of Europe on 24th January, 1975," Feb. 14, 1975, OECD Archives, ENV(75)17. 11.853, 2.

107. Ibid, 3. "Environment Committee: Cooperative Technical Programme to Measure the Long Range Transport of Air Pollutants—Summary Report," Apr. 22, 1977, OECD Archives, (77)23 & Corrigendum, 1–2.

108. Ibid., 5–8.

109. "Committee Room Document No. 1 for the 21st Session of the Committee," Nov. 28, 1977, OECD Archives, ENV—77—DIVERS. E.2147, 1–2.

110. "Million-Dollar Problem."

111. Lean and Mayer, "Britain Moving"; Mayer, "Britain Slashes Research."

112. Pearce, "Research Cuts."

113. For information on Thatcher's statements and stress placed by her administration's Department of Energy on the "uncertainties" of the science behind acid rain, see Beardsley, "Acid Rain"; "Causes of Acid Rain"; "Norway Protests to Thatcher."

114. Milne, "CEGB Hits Back"; Parry, "CEGB 'Continuing Lie'"; UK House of Commons, Parliamentary Debates, *Air Pollution*; "Norway Is Angry."

115. Brynjulf Ottar, *Reduksjon Av Luftforurensningene I Europa. Seminar Om Sur Nedbør, Leangkollen,* Jan. 24–26, 1983, NILU Archives and Library, NILU F 4/83, 12.

116. Sivertsen, *NILU; 40 år i lufta,* 27.

117. Ottar, *Reduksjon Av Luftforurensningene I Europa* (n. 116), 1, 12.

118. Pearce, "Britain Funds Independent Research"; Samstag, "Norway Loses Patience"; Pearce, "Norwegians Protest"; "Norway Protests"; Samstag, "Softly, Softly Plea"; Cramb, "Acid Rain Plea."

119. United Nations, "Protocol to the 1979 Convention."

120. Milne and MacKenzie, "Europe's Worst Polluter"; United Nations, *Environmental Conventions,* 15–20.

121. "Council: Meeting of the Environment Committee at Ministerial Level (Note by the Environment Committee)," Feb. 27, 1974, OECD Archives, /M(74)6(Prov.), 6.

122. OECD, *Legal Aspects of Transfrontier Pollution,* 7.

REFERENCES

Ahrens, C. Donald. *Meteorology Today: An Introduction to Weather, Climate, and the Environment.* 8th ed. Independence, KY: Cengage Learning, 2007.

"Air Pollution by Smoke." *Journal of the American Medical Association* 97, no. 18 (Oct. 31, 1931): 1312–13.

Allix, André. "A propos des brouillards lyonnais. 4. Le brouillard mortel de Liège et les risques pour Lyon." *Les Études Rhodaniennes* 8, no. 3 (1932): 133–44.

"Atmospheric Pollution and Pulmonary Diseases." *British Medical Journal* 1, no. 3658 (Feb. 14, 1931): 277.

Beardsley, Tim. "Acid Rain: UK in the Minority of One." *Nature* 309, no. 5971 (1984): 740–41.

Bell, Michelle L., Devra L. Davis, and Tony Fletcher. "A Retrospective Assessment of Mortality from the London Smog Episode of 1952: The Role of Influenza and Pollution." *Environmental Health Perspectives* 112, no. 1 (Jan. 2004): 6–8.

Bolin, Bert. *Air Pollution Across National Boundaries: The Impact on the Environment of Sulfur in Air and Precipitation, Sweden's Case Study for the United Nations Conference on the Human Environment.* Stockholm: P. A. Norstedt, 1972.

———. "Carl-Gustaf Rossby: The Stockholm Period 1947–1957." *Tellus B* 51, no. 1 (Feb. 1, 1999): 4–12.

"Causes of Acid Rain 'Still Not Understood.'" *Guardian* (London), Aug. 6, 1984, 4.

Cowling, Ellis B. "Acid Precipitation in Historical Perspective." *Environmental Science & Technology* 16, no. 2 (Feb. 1, 1982): 110–23A.

Cramb, Auslan. "Acid Rain Plea by the Scandinavians." *Glasgow Herald*, June 20, 1989, 3.

Dalton, Russell J. *The Green Rainbow: Environmental Groups in Western Europe.* New Haven, CT: Yale University Press, 1994.

Dewey, Scott Hamilton. *Don't Breathe the Air: Air Pollution and US Environmental Politics, 1945–1970.* College Station: Texas A&M University Press, 2000.

Ekbladh, David. *The Great American Mission: Modernization and the Construction of an American World Order.* Princeton, NJ: Princeton University Press, 2009.

Firket, J. "Sur les causes des accidents survenus dans la vallée de la Meuse, lors des brouillards de décembre 1930." *Bulletin de l'Académie Royale Médicine du Belgique* 11, no. 683–741 (1931).

Haldane, J. S. "Atmospheric Pollution and Fogs." *British Medical Journal* 1, no. 3660 (Feb. 28, 1931): 366–67.

Halliday, E. C. "A Historical Review of Atmospheric Pollution." In *Air Pollution*, 11. Geneva: World Health Organization, 1961.

Hesthagen, Trygve, Iver H. Sevaldrud, and Hans M. Berger. "Assessment of Damage to Fish Populations in Norwegian Lakes Due to Acidification." *Ambio* 28, no. 2 (Mar. 1, 1999): 112–17.

Kaijser, Arne. "The Trail from Trail: New Challenges for Historians of Technology." *Technology and Culture* 52, no. 1 (2011): 131–42.

Keyes, D. B. "Pure Air for Our Cities." *Scientific Monthly* 35, no. 5 (Nov. 1932): 427–30.

Krige, John. *American Hegemony and the Postwar Reconstruction of Science in Europe.* Cambridge, MA: MIT Press, 2008.

Launius, Roger D., James Rodger Fleming, and David H. DeVorkin, eds. *Globalizing Polar Science: Reconsidering the International Polar and Geophysical Years.* New York: Palgrave, 2010.

Lean, Geoffrey, and Marek Mayer. "Britain Moving Towards Acid Rain Disaster." *Observer* (London), June 13, 1982, 2.

Long, Bill L. *International Environmental Issues and the OECD, 1950–2000: An Historical Perspective.* Paris: OECD Publishing, 2000.

Mayer, Marek. "Britain Slashes Research on Air Pollution." *New Scientist* (Apr. 29, 1982): 271.

"Million-Dollar Problem—Billion-Dollar Solution?" *Nature* 268, no. 5616 (1977): 89.

Milne, Roger, and Debora MacKenzie. "Europe's Worst Polluter Is Tamed." *New Scientist* (June 23, 1988): 29.

Milne, Seamus. "CEGB Hits Back at Acid Rain Accusers." *Guardian* (London), Oct. 10, 1984, 3.

"Norway Is Angry over Film on Acid Rain." *New Scientist* (Oct. 31, 1985): 13.

"Norway Protests to Thatcher." *New Scientist* (Dec. 12, 1985): 13.

Odén, Svante. "Acidification of Air and Precipitation and Its Consequences on the Natural Environment." Stockholm: Swedish State Natural Science Research Council, 1968.

———. "Nederbördens försurning." *Dagens nyheter*, Oct. 24, 1967.

OECD. *Ad Hoc Meeting on Acidity and Concentration of Sulphate in Rain, May 1969*. Paris: OECD, 1971.

———. *Air Management Problems and Related Technical Studies: Policy Report of the Air Management Research Group*. Paris: OECD, 1972.

———. *Air Pollution in the Iron and Steel Industry*. Special Committee for Iron and Steel. Paris: OECD, 1963.

———. *International Scientific Cooperation*. Paris: OECD, 1965.

———. *Legal Aspects of Transfrontier Pollution*. Paris: OECD, 1977.

———. *Methods of Measuring Air Pollution: Report of the Working Party on Methods of Measuring Air Pollution and Survey Techniques*. Paris: OECD, 1964.

———. *OECD and the Environment*. Paris: OECD, 1976.

———. *The OECD Programme on Long Range Transport of Air Pollutants: Measurements and Findings*. Paris: OECD, 1977.

Ottar, Brynjulf. "An Assessment of the OECD Study on Long Range Transport of Air Pollutants (LRTAP)." *Atmospheric Environment (1967)* 12, no. 1 (1978): 445–54.

———. "The Long Range Transport of Sulfurous Aerosol to Scandinavia." *Annals of the New York Academy of Sciences* 338, no. 1 (1980): 504–14.

———. "Monitoring Long-Range Transport of Air Pollutants: The OECD Study." *Ambio* 5, no. 5–6 (1976): 203–206.

———. "Organization of Long Range Transport of Air Pollution Monitoring in Europe." *Water, Air, and Soil Pollution* 6, no. 2 (1976): 219–29.

Parry, Gareth. "CEGB 'Continuing Lie in Acid Rain Battle.'" *Guardian* (London), Dec. 3, 1984, 2.

Pearce, Fred. "Britain Funds Independent Research on Acid Rain." *New Scientist* (Sept. 8, 1983).

———. "Norwegians Protest over Gag on Research." *New Scientist* (Mar. 20, 1986): 24.

———. "Research Cuts Corrode Britain's Acid Rain Strategy." *New Scientist* (July 15, 1982): 141.

———. "Warning Cones Hoisted as Acid Rain Clouds Gather." *New Scientist* (June 24, 1982): 828.

Regens, James L., and Robert W. Rycroft. *The Acid Rain Controversy*. Pittsburgh: University of Pittsburgh Press, 1988.

Roderick, Hilliard. "The Future Natural Sciences Programme of Unesco." *Nature* 195, no. 4838 (1962): 215–22.

Rodhe, Henning. "Human Impact on the Atmospheric Sulfur Balance." *Tellus B* 51, no. 1 (Feb. 1, 1999): 110–22.

Rossby, Carl-Gustaf. "Current Problems in Meteorology." In *The Atmosphere and the Sea in Motion: Scientific Contributions to the Rossby Memorial Volume*, edited by Bert Bolin, 9–50. New York: Rockefeller Institute Press, 1959.

Samstag, Tony. "Norway Loses Patience with Britain over Acid Rain." *Times* (London), Dec. 9, 1985, 7.

———. "Softly, Softly Plea to Britain on Acid Rain; Norwegian Environment Minister Surlien Visits London." *Times* (London), Mar. 17, 1986, 8.

Schmandt, Jurgen, and Hilliard Roderick. *Acid Rain and Friendly Neighbors: The Policy Dispute between Canada and the United States*. Duke Press Policy Studies. Durham, NC: Duke University Press, 1985.

Sivertsen, Bjarne. *NILU; 40 år i lufta*. Kjeller: Skedsmo, Trykk & Grafisk, 2009.

Sohn, Louis B. "The Stockholm Declaration on the Human Environment." *Harvard International Law Journal* 14, no. 3 (1973): 423–515.

Staples, Amy L. S. *The Birth of Development: How the World Bank, Food and Agriculture Organization, and World Health Organization Have Changed the World, 1945–1965*. Annotated edition. Kent, OH: Kent State University Press, 2006.

Stine, Jeffrey K., and Joel A. Tarr, "At the Intersection of Histories: Technology and the Environment." *Technology and Culture* 39, no. 4 (1998): 601–40.

Talman, C. F. "Death-Dealing Fogs a Scientific Puzzle: Two Aspects of the Strange Effects of the Ice Fog." *New York Times*, Dec. 14, 1930, 6.

ten Houten, J. G., ed. *Air Pollution: Proceedings of the First European Congress on the Influence of Air Pollution on Plants and Animals, Wageningen, April 22 to 27, 1968*. Wageningen: Centre for Agricultural Publishing and Documentation, 1969.

UK House of Commons, Parliamentary Debates. *Air Pollution*. London: Hansard, 1980, cc. 301–4W.

United Nations. *Environmental Conventions: Elaborated under the Auspices of the United Nations Economic Commission for Europe*. New York: United Nations, 1994.

———. "Protocol to the 1979 Convention on Long-Range Transboundary Air Pollution on the Reduction of Sulphur Emissions or Their Transboundary Fluxes by at Least 30 Per Cent." Helsinki. July 8, 1985.

US Congress, Senate. Committee on Commerce. *1972 Survey of Environmental Activities of International Organizations*, Washington, DC: USGPO, 1972, 75–99.

———. Committee on Foreign Relations. *Background Documents Relating to the Organization for Economic Cooperation and Development*. Washington DC: USGPO, 1961.

10

THE TRANSMUTATION OF OZONE IN THE EARLY 1970S

Matthias Dörries

DURING THE EARLY 1970s, public perceptions of ozone underwent a profound transformation. Previously regarded primarily as an annoying result of urban pollution, ozone became a molecule that protected the earth from harmful radiation. This new awareness of the role of the atmosphere in supporting life on Earth was part of a more general trend in which atmospheric scientists transformed disconnected studies of city pollution into a national and international network of work on global atmospheric pollution. Ozone was central to this transformation, as it initiated political and public debates about the use of technologies and substances that apparently affected the ozone layer in the stratosphere and in that way caused possible health effects, like skin cancer due to higher UV radiation.

While the impact of supersonic transport (SST) and chlorofluoro-carbons (CFCs) on the ozone layer were the most prominent publicly debated atmospheric-environmental concerns during the early 1970s, there was a parallel and partly connected debate about how a nuclear war between the superpowers might affect the atmosphere and the ozone layer. Early results from systematic research in the early 1970s pointed to a possibly disastrous impact on the biosphere. Here protecting stratospheric ozone had the potential to serve as a strong cause for disarmament on the political and diplomatic front.

While the history of SST and CFCs has been studied extensively,[1] the debate around the effects of nuclear weapons on ozone in the early 1970s has been neglected. A closer look at that debate will reveal two things: First, perceptions of ozone changed dramatically. Within a military research context, ozone had figured as a strategic tool of warfare; now it increasingly appeared as a "positive" molecule, protecting the Earth not only from harmful radiation, but also from the likelihood of a nuclear war. Ozone was no longer exclusively "bad" ozone, a contaminant, pollutant, and an instrument of destructive warfare but also "good" ozone, a natural component of the atmosphere.[2] Here the omnipresent environmental discourse of the early 1970s crept into scientific publications, and beyond, into the public realm. Second, that debate took place in a considerably more cooperative political environment than did the nuclear winter debate of the 1980s or current climate change discussions. In the early 1970s environmental issues, now taking into account the limited resources of the Earth, were of common interest to politicians, economists, and scientists across the political spectrum in search for rapid answers to pressing societal and economic issues.

Here I trace elements of the evolution of the debate about ozone from the 1950s on, focusing in particular on the early 1970s. I look at decisive transformations of the atmospheric sciences from a disparate set of fields (primarily before 1960) into the leading interdisciplinary edge of global environmental research. During the 1970s, ozone stood in the middle of a multitude of debates, such as how to deal with uncertainty in scientific research, what to do with modeling in the atmospheric sciences, and how to make scientific and military research more transparent and democratic. The stratospheric pollution caused by nuclear bombs became an openly discussed issue, with consequences not only for the careers of researchers but also for their methods and for the language in which they framed their object of research. Two case studies, one on the work of two physicists at Columbia University and one on the link between ozone and disarmament, further elucidate this development.

OZONE AS "BAD OZONE" AND STRATEGIC WEAPON IN THE 1950S AND 1960S

During the early 1950s, ozone emerged as one of the culprits of air pollution.[3] From the 1940s on, the rapidly growing city of Los Angeles had suffered increasingly from smoke and fog, or "smog," as it became fashionable to call it. For a while, confusion and debate reigned about what exactly caused smog, but thanks to the pioneering experimental work of the chemist Arie Jan Haagen-Smit of the California Institute of Technology in Pasadena, it became increasingly clear that sunshine caused photochemical reactions of hydrocarbons and led to a rapid increase in the production of peroxides and ozone.[4] Ozone now started to stand in for smog and such undesirable effects as "decrease in visibility, crop damage, eye irritation, objectionable odor, and rubber deterioration."[5] Ozone and peroxides resulted from such "potentially toxic materials" as nitrogen oxides or hydrocarbons.[6] Hence, the problem was anthropogenic and not a natural phenomenon, in contradiction of earlier suggestions that atmospheric conditions caused stratospheric ozone to descend into Los Angeles. Haagen-Smit challenged these dubious interpretations of the causes of pollution, and struggled with industry, particularly the car industry.[7] He spoke out in favor of "stringent regulations" and "control measures in industry or on the privately owned automobile."[8] Faced with serious pollution, the state of California passed numerous laws to protect its citizens against ozone beginning in the late 1940s, and thus was the first to set worldwide standards for a healthier environment.

While ozone had emerged in Los Angeles as a local contaminant of the troposphere, it was also already known by the 1880s that it occurred naturally in the stratosphere.[9] In 1913 the French scientists Charles Fabry and Henri Buisson took quantitative measurements and were able to attribute ultraviolet absorption in the upper atmosphere to a thin layer of ozone.[10] To gain further insights into the global and vertical distribution of ozone in the atmosphere and its seasonal variations, the International Union of Geodesy and Geophysics (IUGG) set up an International Ozone Commission in 1948, and the International Geophysical Year (IGY) of 1957–58 subsequently led to a vast expansion of an already existing network of so-called Dobson spectrophotometers, invented by the British meteorologist G. M. B. Dobson during the 1920s.[11] Measurements of total levels of ozone were also used for weather forecasting, as they could be linked to the passage of weather systems. Furthermore, in the upper atmosphere, ozone functioned as a tracer, as it had a relatively long lifetime and could therefore reveal global circulation patterns.

With the arrival of atom bombs, missile systems, space exploration, and the IGY, the stratosphere became a new battleground for the Cold War, leading to wild speculations about how it potentially could be modified for military purposes.[12] Knowledge of the circulation of the atmosphere and its chemical processes was still extremely limited, but in the early 1960s there was already talk of modifying (voluntarily or involuntarily) the strato-spheric ozone layer with specific chemical substances. In a 1962 presentation, "On the Possibilities of Climate Control," Harry Wexler, head of research in the US Weather Bureau and a leading American meteorological expert working closely with the military, worried about "atmospheric pollution" in the stratosphere by burning rocket fuel. The release of chlorine or bromine from the fuel oxidizers had the potential to "destroy naturally occurring atmospheric ozone and open up a 'hole,' admitting passage of harmful ul-traviolet radiation to the lower atmosphere."[13] Wexler "was clearly interested in both inadvertent climatic effects—such as might be created by industrial emissions, rocket exhaust gases, or space experiments gone awry—and pur-poseful interventions, whether peaceful or done with hostile intent."[14] He picked up John von Neumann's discussion of climatological warfare in the 1950s. James Rodger Fleming has looked at the correspondence between Harry Wexler and the Caltech chemist Oliver Wulf between December 1961 and April 1962, when Wulf suggested chlorine and bromine as "deozonizers." Wexler scribbled down notes that included speculative calculations about how much bromine would be needed to destroy all stratospheric ozone in the polar region and near the equator; however, these detailed reflections and calculations were never published, as Wexler died in August 1962.[15]

While the chemical mechanisms seemed to have been lost from sight, the idea of manipulating the ozone layer nevertheless lived on. Following Wexler's ideas, Gordon J. F. MacDonald, a geophysicist at the University of California at Los Angeles, asserted in 1968 that creating a "temporary 'hole'" in the atmosphere could be fatal to all forms of life. MacDonald engaged in speculation: "Among future means of obtaining national objectives by force, one possibility hinges on man's ability to control and manipulate the environment of his planet. When achieved, this power over his environment will provide man with a new force capable of doing great and indiscriminate damage. Our present primitive understanding of deliberate environmental change makes it difficult to imagine a world in which geophysical warfare is practised."[16] While MacDonald warned of future threats to the environ-ment, he did so more in the spirit of awe of human capabilities than fear for the future of the Earth that motivated environmentalists. A series of mili-tary studies had addressed all kinds of atmospheric issues, but this research was mostly classified or appeared in laboratory reports that were not easily accessible, and never coalesced into a coherent or ambitious research pro-gram. Military research on the atmosphere up to the 1970s remained highly speculative and dispersed, driven by the moment. Often these reports did not mention the original motivation for doing a certain type of research. Ozone in these publications served in a manifold way, as an indicator of patterns of atmospheric circulation, as a possible strategic weapon, or as a way to detect missiles.[17] As long as fear of the enemy still dominated over fear of human self-destruction, ozone was of strategic importance. As such it was both a threat *and* an opportunity.

Ozone as "Good Ozone" in the 1970s

In a series of dramatic turns, atmospheric science, and, with it ozone research, changed within a few years in the late 1960s and early 1970s. The turns happened at several levels: at the scientific level, thanks to a better understanding of the chemical processes and increased atmospheric modeling, done either in the laboratory or with the help of computers; at the political level, exemplified by the creation of new environmental institutions and investigations of the environmental impact of new technologies; at the military level, exemplified by increasingly public debates on the environmental and human consequences of military weapons, particularly nuclear bombs; and finally, at a cultural level, exemplified by a new way of talking and of framing scientific issues.

Studies of the atmosphere and ozone expanded spectacularly in the early 1970s, with President Nixon's decision in September 1969 to go ahead with building the supersonic transport. SSTs were supposed to cruise in the stratosphere, which meant that their exhaust gases would remain trapped there for several years, potentially causing environmental damage.[18] At the time, atmospheric scientists had little knowledge about the exact chemical and physical processes involved. However, Congress and the emerging ecological and environmental movements pressed for answers. In the words of British climatologist Hubert H. Lamb, there was "a widespread growth of public concern over what man is doing to his environment."[19] In fact in the United States, the National Environmental Policy Act (NEPA) of 1969 had led to the creation of the Council of Environmental Quality, which was to report annually to the president. Nixon wrote in the foreword to the first report of 1970, "It is also vital that our entire society develop a new understanding and a new awareness of man's relation to his environment—what might be called 'environmental literacy.'"[20]

Pollution from human sources, either in the troposphere or in the stratosphere, moved into the midst of public concerns and debates. In the troposphere ozone figured as one of the seven key pollutants in the 1970 revision of the Clean Air Act, while in the stratosphere ozone was indispensable to protect human beings from dangerous radiation. However, little was known about the exact chain of chemical processes or about global and vertical distribution of ozone. The upcoming United Nations Conference on the Human Environment in Stockholm in 1972 provided a welcome focus for pushing a new research agenda for atmospheric research for the next decade. In preparation for this meeting, leading American scientists were involved in writing the Study of Critical Environmental Problems (SCEP) report of 1970 on "Man's Impact on the Global Environment," and the "Study of Man's Impact on Climate" report of 1971, which laid out a program for "a comprehensive investigation of global environmental problems" for "the international scientific community."[21]

In fact, within a few years atmospheric chemistry and modeling underwent a complete transformation, exemplified by a series of pathbreaking and Nobel-winning discoveries, such as Paul Crutzen's work on the interaction of NO_x and ozone in 1970, and the work by F. Sherwood Rowland and Mario Molina in 1974 on CFCs and ozone depletion in the stratosphere, to name only the most visible.[22] Furthermore, SST had also led, at the instigation of the Department of Transportation, to the $20 million establishment of the Climate Impact Assessment Program (CIAP), a program that generously funded a flood of stratospheric research between 1971 and 1975.[23] As this kind of research was supposed to provide rapid answers to pressing political and economic issues, modeling increasingly substituted merely empirical research methods in atmospheric science, challenging traditional practices.

MILITARY RESEARCH ON OZONE AND NEW PUBLIC SCRUTINY IN THE 1970S

In the early 1970s, atmospheric studies sponsored by the military, ranging from missile detection to the effects of nuclear explosions, and weather and climate modification were revisited in the light of the urgency of action to avoid possible environmental and economic damage on a national and global scale. Previous military research was dragged into public and scientific debate and underwent new, critical scrutiny. Scientists working for the military or for military think tanks were faced with a number of probing scientific and political questions. In the following, I look at two debates around ozone and nuclear bombs in the 1970s. First is the scientific and political debate around the work in 1972–73 by two Columbia physics professors who were the first to make statements about the impact of nuclear bombs on ozone. This case will serve to show to what extent military research on ozone had moved into the public realm and the extent to which new ways of doing atmospheric science replaced the modes of the 1960s. Second, I will look at how the widespread public debate on ozone spilled over into the debate about disarmament in 1974, leading to political conflict about military secrecy over ozone studies.

Studying Ozone Depletion Due to Nuclear Bombs in the 1970s

The rise of environmental thinking, and the ozone debate around SST, dragged military research out of its usual cover, as studies on stratospheric pollution caused by nuclear weapons promised to provide insight into the possible effects of an SST fleet. In the early 1970s confusion over the effects of the large atmospheric nuclear explosions of the early 1960s on the ozone level reigned. In the 1960s there had been some speculation that these explosions affected ozone, but given the complexities of chemical reactions in the atmosphere, a firm grasp of the relevant mechanisms remained elusive. Paul Crutzen's seminal paper of 1970 on ozone depletion due to nitrous oxide molecules (NO_x), gave research a new direction, as it argued that NO_x acted as a catalyst, and thus did not diminish. This meant that NO_x could lead to large ozone depletions, especially as it remained in the stratosphere for several years. While the production of massive amounts of NO_x due to nuclear explosions was without doubt, the degree of ozone depletion remained uncertain.[24] Since atmospheric nuclear tests had stopped in 1963 after the moratorium between the two superpowers, researchers had to rely on earlier measurements. Systematic measurements of ozone had been collected by the World Ozone Data Centre in Toronto, established in 1961 at the instigation of the World Meteorological Organization.[25] Attempts to use a satellite, Nimbus IV, to make direct ozone observations following French and Chinese atomic explosions conducted in 1970 were inconclusive, since the sensors failed to detect significant amounts of ozone.[26]

Faced with the pressing issue of providing a reasonable account of how SSTs would affect ozone in the stratosphere, scientists searched for analogies and went back to the nuclear explosions of the 1960s, which produced substantial amounts of NO_x, similar to what SSTs might produce, and possibly had led to ozone depletion. The first paper to address the issue of NO_x and nuclear weapons emerged in the summer of 1972 from the so-called Jason Division, an elitist and discreet group of roughly forty scientists (mostly theoretical physicists) who had met regularly for several weeks in the summers since 1958 under the auspices of the Institute for Defense Analyses (IDA). These scientists (among them Gordon J. F. MacDonald, mentioned above) debated and worked on all matters of military relevance and had access to classified materials.[27] The paper, which first appeared as an IDA in-house study in 1972 and soon after in the *Journal of Geophysical Research*, was written by two Columbia University physicists, Henry M. Foley and Malvin A. Ruderman.[28]

Foley's and Ruderman's paper was a typical think-tank armchair paper, in that it did not do any new research, but relied on reviewing previous studies, enhanced with some additional knowledge taken from classified material. Foley and Ruderman confirmed the large NO_x production during the high-megaton nuclear explosions of the early 1960s and saw them as similar to the NO_x production of a fleet of 500 SSTs over a year. Relying on multiple series of global and local ozone measurements, they could see "no evidence of a large ozone decrease in the months following the test explosions."[29]

There was a political as well as a scientific reaction towards Foley's and Ruderman's work, done in the context of strong anti–Vietnam War actions on the campus of Columbia University. Faculty members had occupied the physics department, accusing five physicists of doing military consultancy for the Vietnam War and of being involved with the Jason Division. Scientists who had worked on military topics became part of an open public and political debate. In their own defense, Foley and Ruderman published a full-page advertisement in the *Columbia Daily Spectator* of April 28, 1972, in which they defended their Jason work and justified their activities by saying that their involvement with the IDA served the purpose of a more open society, as otherwise "military research would go on, but in a much more closed society than exists now."[30] Frank Baldwin, who had obtained a PhD in Asian studies at Columbia in 1969 and was active in the anti–Vietnam War movement, struck back at this assertion, saying, "The Foley-Ruderman thesis is a familiar one: it is preferable to work within an organization to influence policy. That may work for some questions, the SST perhaps, but the requirements of team membership also include silence on the really critical problems such as the war. The argument has been heard so many times over the last eight years from tired, embarrassed liberals retreating from their vineyards of death at the Pentagon and the State Department that few take it seriously any more."[31] For Baldwin, Foley's and Ruderman's self-ascribed ideal of the independent scholar was an impossibility, especially in the context of the ongoing Vietnam War. Though Foley and Ruderman seem not to have done work related to the Vietnam War, they found themselves in the middle of a political storm. Furthermore, their involvement with the Jasons meant additional political trouble, as their study indirectly supported the SST project. As Walter Sullivan pointed out in an article in the *New York*

Times, several IDA members were involved in the assessment of the SST and the program manager, Alan J. Grobecker, was a former member of the IDA.[32] Atmospheric research could no longer be done outside public scrutiny and now stood in the middle of cultural and political debates about US engagement in the Vietnam War and environmental and economic ones on SST. Dragged into the public limelight, Foley and Ruderman were unprepared.

More criticism was to follow, this time from fellow scientists. Julius S. Chang of the Lawrence Livermore National Laboratory (LLNL) strongly criticized their use of data. Chang, trained in physics and mathematics, had earned a PhD in applied mathematics and statistics from the State University of New York, Stony Brook, in 1971. Starting out with some classified work in computational physics and modeling, he became heavily involved in CIAP and transformed into an expert on stratospheric chemistry and transport modeling. He accused Foley and Ruderman of having done a "very cursory examination of the available ozone data," of excessive simplification of the physical processes involved, and of neglect of the differences between the immediate injection of NO_x by nuclear weapons and the rather continuous injections by SST. According to Chang, Foley and Ruderman had had "expectations of an unrealistically large perturbation in the O_3 observations," and then unsurprisingly did not find any major changes.[33] Based on some preliminary and rudimentary calculations, Chang suggested a 15 percent reduction in total ozone. A few months later, in May 1973, in a more elaborate study using a one-dimensional vertical transport model with coupled chemical kinetics, he calculated a 4 percent reduction of total ozone in the northern hemisphere, similar to natural variations. Therefore he ultimately discarded the influence of nuclear weapon explosions on ozone in the stratosphere.[34] Still, Chang's criticism made it clear that Foley's and Ruderman's armchair physics could no longer be the standard, and that modeling and intimate knowledge of the chemical issues involved were the new standard of atmospheric science.

Other research groups had meanwhile picked up the issue of nuclear weapons. In the United States, Harold S. Johnston, from the department of chemistry of the University of California at Berkeley, the author (together with Gary Whitten and John Birks) of the widely debated 1971 study of possible major reductions of ozone by NO_x from SSTs, critically examined Foley's and Ruderman's article.[35] Because no direct measurements of NO_x had been done during the tests in the early 1960s, Johnston et al. relied on analogies to studies of radioactive particles to draw analogies for the distribution of NO_x. They studied the distribution of NO_x in the stratosphere and argued that NO_x produced by nuclear bombs was located in the lower stratosphere and in the higher northern latitudes. From this, they concluded that only a reduction of a few percent seemed likely, thus reducing previous expectations of much larger ozone depletion. However, they interpreted the available data quite differently than did Foley and Ruderman. They regarded the 5 percent increase of total ozone after the moratorium between 1963 and 1970, when no atmospheric tests took place, as a confirmation of their hypothesis that nuclear bombs depleted ozone.[36]

Walter Sullivan, a reporter for the *New York Times*, perceptively noted in 1972 that multiple uncertainties made any study "strongly influenced by . . . personal views as to the social, political or economic merits" of SST.[37] For example, P. Goldsmith of the British Meteorological Office, an ardent defender of the SST, dismissed Johnston's use of data as statistically insignificant, given the large daily and annual fluctuations, and saw no detectable changes in the total ozone concentration. Goldsmith also differentiated between laboratory studies on NO_x and ozone (as done by Johnston et al.) and processes in the stratosphere, which might be affected by a number of additional factors, like solar radiation, stratospheric circulation, refinement of chemical rate coefficients, and others.[38]

To summarize the different methodological options and their respective criticisms: Armchair physics, as was done in think tanks, was no longer acceptable to the public and scientists, as it was not based on original research but only on the re-analysis of existing data, resulting in speculation that was no longer acceptable to environmentally conscious scientists. Atmospheric modeling in the early 1970s was still in its infancy, suffering from a lack of sufficient data and of access to sufficient computer power, and therefore could provide rudimentary hints at most, "classical" meteorologists judged. Finally, laboratory experimentation was able to produce atmospheric phenomena under artificial conditions, but critics charged that it remained a far cry from real-world conditions.

Uncertainty ruled, and unsurprisingly then, the matter remained unresolved. Still, there was public and political pressure for answers. CIAP had been launched with a strict deadline of final results within a few years. It was a "crash program," which left little time for long-lasting and statistically relevant measurements.[39] As a review article of 1975 noticed, this did not leave any other choice but laboratory experimentation or modeling to provide at least partial answers in time scales of a few years: "In view of the short time scale of the program, the principal method of attack on the problem has consisted of measurements of ambient stratospheric NO_x and other potentially relevant species supplemented by extensive computer modeling. To augment our confidence in these predictions, any kind of partial experimental simulation is worthy of investigation."[40]

Ozone and Disarmament in 1974

The military was not exempt from political pressure to come up with rapid answers. In fact, a possible destruction of the ozone layer due to nuclear bombs became the worry within the military in the early 1970s. At first the military tried to deal with the issue in the usual secret manner. However, with ozone and nuclear bombs increasingly moving to the forefront of public debate, it ultimately was forced into addressing the issue in a public statement in 1974, as a result of tensions within military circles in Washington. The military worry had its origin in an article of 1971 on SST by Harold Johnston, an authority in the field of atmospheric science who argued that "if concentrations of NO and NO_2 are increased in the stratosphere by the amounts accepted by the SCEP report and by governmental agencies, then there would be a major reduction in the O_3 shield (by about a factor of 2 even when allowance is made for less NO_x emission than SCEP used)."[41] When this reasoning was applied to the effects of nuclear weapons, a nuclear war could also possibly lead to a reduction of 50 percent of ozone in the stratosphere,[42] with possibly disastrous long-term consequences for the food chain. This seems to have been the driving force behind a flood of studies on nuclear weapons and their effects on stratospheric ozone. Beside the ones examined above, there was also a series of classified AEC studies that pointed this direction. Since the issue was of importance for military and political strategy, the DoD pushed for further research, and the first results seemed to indicate strong ozone depletion for the case of a nuclear war between the superpowers. However, this research remained classified.

Given the political importance of the issue, Fred C. Iklé, director of the US Arms Control and Disarmament Agency (ACDA), took action on April 4, 1974, and asked the president of the National Academy of Sciences (NAS) to assess the likely long-term, worldwide effects on the environment and ecology of nuclear weapons attacks. Iklé argued in his letter that only the NAS could address the "many interrelated physical aspects" of the problem.[43] No explicit mention of ozone was made in the letter, but it was the driving force behind Iklé's actions. Iklé seemed to have been displeased by the DoD's decision to classify its research. An article by John W. Finney in the *New York Times* mentioned "complaints [by the ACDA] that the Defense Department has refused to make available its studies on the possible effects of nuclear explosions on the ozone layer."[44] Iklé's actions, driven by a sense of urgency to find broad scientific consensus, displeased the Pentagon, which now saw itself under pressure to come out with its own interpretation.

The AEC considered a press release and was in discussion with the directors of the military laboratories. A declassified fax of May 2, 1974, from the director of LLNL, Roger E. Batzel, to AEC officials and directors of other military laboratories, sheds light on discussions within the military research community. Batzel had made changes to the original version, as he wished to phrase the issue in a somewhat less speculative, less spectacular, and "somewhat truer perspective" to "retain credibility." Nevertheless, he feared a public storm, stating, "However the final release is worded, it seems likely to prompt press inquiries and considerable attention. We believe it would be useful if advance arrangements are made to have qualified scientists available to help provide answers."[45] While Batzel put his hope in "qualified scientists," some members of the scientific community beyond the military laboratories drew attention to the issue, and talked about how the latest research would affect the disarmament debate. John Hampson, an atmospheric researcher who had worked for the Canadian Armament Research and Development Establishment in the 1960s, deplored the apparent lack of studies on the disastrous consequences of nuclear war, particularly on the effects of ozone, in light of Johnston's work, and pleaded for the inclusion of this topic in UN deliberations to limit nuclear proliferation. Hampson was not soft on his colleagues: "The majority of the scientific community fail to recognise the potential danger from a change in the chemical envelope of the Earth. They thus fail to consider the possibility of defining a limit to nuclear activity, which can be universally agreed upon and could set the stage for arms control. . . . All the parameters involved must be studied, and all potential, alterative models of stratospheric chemical behaviour must be included in order to avoid a superficial interpretation which fits the selected facts visible from the fraction of the iceberg that can be seen at present."[46] Ozone got even more attention in 1974, as Mario J. Molina and F. Sherwood Rowland of the University of California at Irvine were writing their seminal article on CFCs and ozone that appeared on June 28, 1974.[47]

The tension between Iklé at the ACDA and the DoD came to the forefront in the fall of 1974, when Iklé addressed the issue of disarmament before the Council of Foreign Relations in a talk entitled "Nuclear Disarmament without Secrecy." This talk was extraordinary in the sense that Iklé stressed uncertainty and the "lack of real knowledge" among those who spoke in an "imaginary order" of megatons, missiles, first strike, or mutual deterrence:

For those of you who have not followed this macabre branch of science closely, I have important news: We are not only unable to express the human meaning of nuclear war—the only meaning that matters—we are also unable to express the full range of physical effects of nuclear warfare, let alone to calculate these effects. Why is this so? Because the damage from nuclear explosions to the fabric of nature and the sphere of living things cascades from one effect to another in ways too complex for our scientists to predict. Indeed, the more we know, the more we know how little we know.[48]

Iklé's actions and his encouragement to shed secrecy and to implicate the NAS may be seen as a way to keep control of the public debate. However, his strong words and energetic pleading for an open discussion about nuclear weapons within the United States and arms reduction negotiations between the superpowers displeased the DoD, which reacted to Iklé and newspaper reports about his presentation with its own press release. It announced that AEC scientists had calculated that a nuclear war could potentially lead to a 50 percent reduction of ozone in the stratosphere with "serious consequences for all nations, including the attacker." Nevertheless, the press release played down the issue, by stressing the uncertainty of the studies and by comparing these changes to living closer to the equator, where ozone protection was lower. It also addressed possible climatic changes due to ozone depletion. It contained numerous qualifying statements, but the issue was nevertheless on the table.

According to reporter John Finney, the Pentagon's assessment was considerably "less ominous in its implications" than the ACDA. The claim was, more or less, that even with ozone depletion in the temperate areas, life would still go on like it does, with less ozone concentration in the tropical areas. This was in stark contrast to Iklé, who had pointed out effects that could possibly affect the food chain, thus shattering "the ecological structure that permits man to remain alive on this planet."[49] Iklé's message appeared in 1975 in the *Bulletin of the Atomic Scientists* along with more information about ozone and a stunning image (see fig. 10.1).

The National Academy report titled *Long-Term Worldwide Effects of Multiple Nuclear-Weapons Detonations* was released on August 12, 1975. It was based on a workshop in January 1975 attended by fifty-six scientists, including Chang. The part on atmospheric effects brought together eleven experts from universities, the IDA, Department of Transportation, the LLNL, the National Oceanic and Atmospheric Administration, and the National Center for Atmospheric Research (NCAR). Concerning ozone, the report ac-

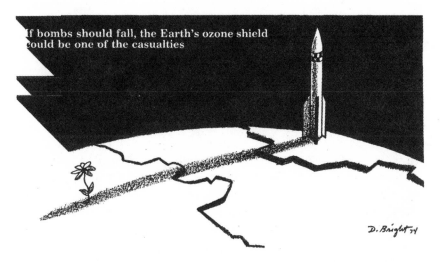

If bombs should fall, the Earth's ozone shield
could be one of the casualties

D. Bright 74

FIG. 10.1. Image linking nuclear weapons, damage to the ozone shield, and damage to the biosphere. SOURCE: Fred C. Iklé, "The Nuclear World of Nuclear Megatonnage," *Bulletin of the Atomic Scientists* 1 (1975), 21. Reprinted with permission from SAGE Publications, Ltd.

knowledged that one-dimensional models showed ozone depletion following a nuclear war of between 30 and 70 percent in the Northern Hemisphere and 20 to 30 percent in the Southern Hemisphere, lasting for two to four years, with possible worldwide effects on climate, crop production, mutagenesis of pathogenic viruses and microorganisms, and other effects. However, it also stressed that these effects would be transitory, disappearing after a few years. It emphasized that because of disturbances in the stratosphere (such as the quasi-biennial oscillations), observational data of the early 1960s could not distinguish signal from noise, and thus could not confirm or refute model calculations. It also discussed the possibility that ozone depletion might reduce global average surface temperature.[50] The ACDA brought out its own version of the results. In the foreword, Iklé took the occasion again to emphasize the "human perspective" and to emphasize the urgent need for arms reductions. He argued that the numerous uncertainties for the Earth were a further deterrent to the use of nuclear weapons, as the "aggressor country might suffer serious physiological, economic, and environmental effects even without a nuclear response by the country attacked."[51]

Toward the end of the 1970s the issue of the effects of nuclear bombs on ozone vanished from the horizon of scientists, politicians, and the public. There was a trend in nuclear weapons toward multiple and smaller warheads of less than one megaton, far lower yields than earlier high-megaton tests. Smaller weapons exploding near the ground would in all likelihood, it was argued, not affect the stratosphere, and hence would not lead to ozone depletion. The Senate Committee on Foreign Relations asked the Office of Technology Assessment (OTA) "to examine the effects of nuclear war on the populations and economies of the United States and the Soviet Union." The composition of the advisory panel was quite different from the NAS study, composed of retired generals and directors of military laboratories, think tank directors (for example, Hudson), professors of genetics, biology, religion, and electrical engineering, complemented by a Soviet specialist and the director of the Emergency Preparedness Project. Cecil Leith of the NCAR represented the atmospheric sciences. While the OTA report of 1979 grudgingly confirmed the possibility of disastrous environmental destruction due to damage to the ozone shield, it distanced itself from the "alarmist" tone of the earlier NAS report.[52]

In the end, after several years of intense work, in the 1970s little that was definitive could be said about ozone depletion, whether caused by SST or by nuclear bombs, with the exception that effects would seem in all likelihood less severe than had been originally inferred. For some, Johnston and Crutzen had been too alarmist; for others, they had drawn attention to the fragility of the Earth's atmosphere. The ozone debate therefore also left bystanders with the impression that scientists, especially those working with modeling and scenarios, tended to overinflate their cause at the expense of a more patient and solidly grounded observational science. When nuclear winter took the stage in the early 1980s, there may have been a déjà-vu syndrome for some, and in fact the debate followed lines similar to the ozone debate. The evolution of debates also points to the speed of transformations within the atmospheric disciplines in the exciting years of the 1970s and 1980s, when surprises were not uncommon and practices and results could change unexpectedly within a few months or years. This is still true today. Recently some atmospheric researchers (among them Richard P. Turco, who had already worked on ozone and nuclear bombs and was a key figure in the nuclear winter theory) revisited the contested issue of nuclear war and ozone depletion and argued that even a regional nuclear conflict would be enough to lead to a "massive global ozone loss."[53]

Ozone played a pivotal role in the new environmental discourses and challenges that came to the forefront in the early 1970s. The new environmental consciousness working as an umbrella for disparate ways of thinking led the actors to frame ozone issues in new terms. Here I can only point to a few examples. While Foley and Ruderman had stated the problem neutrally as "ozone reduction" or "ozone decrease,"[54] the standard term was soon to be "ozone depletion," having the much value-laden connotation of decline and of limits of resources. By the middle of the 1970s this term was generally accepted and in widespread use. To emphasize the protective function of ozone for the Earth's atmosphere, atmospheric scientists like Johnston or Crutzen used the more polarizing term "ozone shield," which replaced the more neutral "ozone layer." Others, like Hampson, spoke of an "ozone screen."[55] A wider and more systematic analysis of the uses of words and terms would probably allow for refinement of the reorganizing and emerging frontlines that would ultimately solidify during the 1980s in the nuclear winter debate and later on in the more recent confrontational debates over anthropogenic climate change.

The new environmental consciousness also reshaped scientific questions. Ozone posed concrete environmental problems of high complexity and public interest, as it confronted scientists, politicians, and the military alike with tricky problems of the distribution and transformation of chemical substances in the atmosphere. As results often remained unresolved and open to revision, uncertainty in atmospheric research became a new focal point for public discussions. Military studies, such as the one by Foley and Ruderman, had already been full of uncertainties, but this had not been a problem for the authors as long as they discussed their research with peers in closed sessions. Foley's and Ruderman's best estimate did not foresee damage to the ozone layer. However, once such studies came under public and scientific scrutiny, the self-assured and predictive character of this type of study came increasingly into doubt and competing interpretations based on alternative investigative methods flourished and put potential severe environmental impacts back on the table.

In the absence and unlikeliness of rapid certitude, acknowledging general uncertainty became a way of forging broad consensus. When Iklé stressed the incertitude of atmospheric research, he abandoned belief in the easy technocratic solutions peddled by physicists since the 1950s. At the same time, he inaugurated a line of argument that would come to be appropriated over the next decade by an established guard of cold warriors and physicists to express skepticism toward the theories of nuclear winter or anthropogenic climate change. This line of argumentation also dismissed modeling as a guide toward more certainty. Here ozone inaugurated a clash between those who argued strongly against modeling and those who expressed trust in the power of modeling in the future.

This is exemplified in a review article from 1975. Ernest Bauer of the IDA and Forrest R. Gilmore of the think tank R&D Associates argued that, in contrast to well-justified predictions, the "existing atmospheric data do not provide a statistically significant demonstration of the catalytic destruction of ozone by oxides of nitrogen."[56] Ambitious younger scientists, who contributed to spectacular advances in the field of atmospheric sciences, especially in atmospheric chemistry, dealt differently with the issue of uncertainty, believing they had good reasons for increasing confidence in their models and atmospheric modeling in general. Not that they believed naïvely in their modeling, but they expected a clear trend toward higher certitude and confidence in the long term and saw themselves as pioneers. For example, R. C. Whitten, W. J. Borucki of NASA-Ames, and R. P. Turco of R&D Associates submitted a paper to *Nature* in which they confirmed Hampson's fear that the Earth's ozone shield might be destroyed by nuclear explosions. Their graphs showed possible ozone depletions from 30 percent to up to 70 percent for various cases.[57] For Chang, the second half of the 1970s was characterized by "significant changes" in the "formulation of one-dimensional models of the stratosphere, and in the experimental values of chemical reaction rate constants used as model input." Furthermore, "substantially more analysis of the ozone record" had been carried out.[58] Walter Sullivan, writing about a CIAP meeting on SST impact on ozone, saw two basic strategies used by atmospheric researchers: "One school favors strong emphasis on simulation of stratospheric processes, including the effect of supersonic transport exhausts on the mammoth computers used for 'modeling' the earth's weather. The other school believes that a realistic answer can only be obtained by measurements and tests within the stratosphere itself."[59]

Despite these methodological differences, despite widespread doubts in Johnston's and Crutzen's "grim" picture, and despite value judgments that informed debates, these different lines of practices and argumentation in ozone debates had not yet hardened into clearly distinctive groups during the early 1970s. Political and scientific alliances were still in a flux of reorganization at the time. Rather, the hardened frontline of the ozone debate of the 1970s was primarily between scientific research on the one side and business interests (such as the supersonic industry) on the other, but much less among scientists themselves or scientists and politicians. Altogether, the ozone debates of the 1970s took place in a more cooperative spirit than the nuclear winter debate a decade later.[60] Ozone research and atmospheric research profited from the favorable political climate for environmental issues under both Republican and Democratic presidents Nixon, Ford, and Carter, which gave research on chemical substances in the atmosphere new status and drive.

NOTES

1. See, for example Dotto and Schiff, *Ozone Wars*; Christie, *Ozone Layer*; and Conway, *Atmospheric Science at NASA*.

2. Crutzen and Ramanathan, "Ascent of Atmospheric Sciences," 301.

3. See Brimblecombe, this volume.

4. Haagen-Smit, "Chemistry and Physiology of Los Angeles Smog," 1342–46.

5. Ibid., 1342.

6. Ibid., 1346.

7. See Dunn and Johnson, this volume.

8. Haagen-Smit, "Atmospheric Reactions of Air Pollutants," 70A.

9. Hartley, "On the Probable Absorption," 268.

10. Fabry and Buisson, "L'absorption de l'ultraviolet," 196–206.

11. Dobson, "Forty Years' Research," 387–405.

12. See Jessee, this volume.

13. Wexler, quoted in Fleming, *Fixing the Sky*, 217.

14. Fleming, *Fixing the Sky*, 218.

15. Ibid., 220–22.

16. MacDonald, "How to Wreck the Environment," 181.

17. Hunt, "Need for a Modified Photochemical Theory," 88–95; Hampson, "Photolysis of Wet Ozone," 215–34.

18. See Jessee, this volume.

19. Lamb, "Critical Problem," 53.

20. Nixon, "President's Message," vii.

21. Carter, "Global Environment," 660.

22. Crutzen, "Influence of Nitrogen Oxides," 320–25; Molina and Rowland, "Stratospheric Sink," 810–12.

23. Conway, *Atmospheric Science*, 133–39.

24. Crutzen, "Influence of Nitrogen Oxides," 320–25.

25. World Ozone and Ultraviolet Radiation Centre, accessed Jan. 30, 2013, http://www.woudc.org.

26. Miller et al., "Nuclear Weapons Tests."

27. For an introduction to this group, see Finkbeiner, *Jasons*.

28. Foley and Ruderman, "Stratospheric NO Production," 4441–50.

29. Ibid., 4448.

30. Foley and Ruderman, "Statement," 9.

31. Baldwin, "Jason Project," 5. For a contemporary look at the issues see, for example, Nelkin, *University and Military Research.*

32. Sullivan, "Scientists Plan Series of Tests."

33. Chang, *Comments on the Possible Effect of NO$_x$ Injection*, 1–2.

34. Chang and Duewer, *On the Possible Effect of NO$_x$ Injection.*

35. Johnston, "Reduction of Stratospheric Ozone," 517–522.

36. Johnston, Whitten, and Birks, "Effect of Nuclear Explosions," 6107–35.

37. Sullivan, "Scientists Plan Series of Tests."

38. Goldsmith et al., "Nitrogen Oxides," 545–51. Goldsmith's article was published in *Nature*, which, under John Maddox, had taken a pro-SST point of view.

39. Sullivan, "Scientists Plan Series of Tests."

40. Bauer and Gilmore, "Effect of Atmospheric Nuclear Explosions," 451.

41. Johnston, "Reduction of Stratospheric Ozone," 522.

42. Batzel, *War Climate Study.*

43. Iklé, letter to Philip Handler, Apr. 4, 1974.

44. Finney, "Pentagon Replies."

45. Batzel, *War Climate Study.*

46. Hampson, "Photochemical War on the Atmosphere," 190–91.

47. Molina and Rowland, "Stratospheric Sink."

48. Iklé, *Nuclear Disarmament*, 3.

49. Finney, "Pentagon Replies."

50. US National Research Council, *Long-Term Worldwide Effects.*

51. Iklé, "Foreword."

52. US Congress Office of Technology Assessment, *Effects of Nuclear War*, foreword, 9.

53. Mills et al., "Massive Global Ozone Loss," 5307–12.

54. Foley and Ruderman, "Stratospheric NO Production," 4441, 4448.

55. Johnston, "Reduction of Stratospheric Ozone," 522; Crutzen, "SSTs," 41–51; Hampson, "Photochemical War on the Atmosphere," 190.

56. Bauer and Gilmore, "Effect of Atmospheric Nuclear Explosions," 451.

57. Whitten, Borucki, and Turco, "Possible Ozone Depletions," 38–39.

58. Chang, Duewer, and Wuebbles, "Atmospheric Nuclear Tests," 1755.

59. Sullivan, "Scientists Plan Series of Tests."

60. Dörries, "Politics of Atmospheric Sciences."

REFERENCES

Baldwin, Frank. "The Jason Project: Academic Freedom and Moral Responsibility." *Bulletin of Concerned Asian Scholars* 5 (1973): 1–12; reprinted from *Christianity and Crisis* (Sept. 18, 1972, and Dec. 11, 1972).

Batzel, Roger E. *War Climate Study—TWX.* UCRL-ID-126117. Livermore, CA: Lawrence Livermore National Laboratory, 1974.

Bauer, E., and F. R. Gilmore. "Effect of Atmospheric Nuclear Explosions on Total Ozone." *Reviews of Geophysics* 13 (1975): 451–58.

Carter, Luther J. "The Global Environment: M.I.T. Study Looks for Danger Signs." *Science* 169 (1970): 660–62.

Chang, Julius S. *Comments on the Possible Effect of NO_x Injection in the Stratosphere Due to Atmospheric Nuclear Weapons Tests.* UCRL-74425. Livermore, CA: Lawrence Livermore Laboratory, 1973.

Chang, Julius S., and W. H. Duewer. *On the Possible Effect of NO_x Injection in the Stratosphere due to Past Atmospheric Nuclear Weapons Tests.* UCRL-74480. Livermore, CA: Lawrence Livermore Laboratory, 1973.

Chang, Julius S., William H. Duewer, and Donald J. Wuebbles. "Atmospheric Nuclear Tests of the 1950s and 1960s: A Possible Test of Ozone Depletion Theories." *Journal of Geophysical Research* 84 (1979): 1755–65.

Christie, Maureen. *The Ozone Layer: A Philosophy of Science Perspective.* Cambridge, UK: Cambridge University Press, 2001.

Conway, Erik M. *Atmospheric Science at NASA: A History.* Baltimore: Johns Hopkins University Press, 2008.

Crutzen, Paul J. "The Influence of Nitrogen Oxides on the Atmospheric Ozone Content." *Quarterly Journal of the Royal Meteorological Society* 96 (1970): 320–25.

———. "SSTs—A Threat to the Earth's Ozone Shield." *Ambio* 1 (1972): 41–51.

Crutzen, Paul J., and Veerabhadran Ramanathan. "The Ascent of Atmospheric Sciences." *Science* 290 (2000): 299–304.

Dobson, G. M. B. "Forty Years' Research on Atmospheric Ozone at Oxford: A History." *Applied Optics* 7 (1968): 387–405.

Dörries, Matthias. "The Politics of Atmospheric Sciences: 'Nuclear Winter' and Global Climate Change." *Osiris* 26 (2011): 198–223.

Dotto, Lydia, and Harold Schiff. *The Ozone Wars.* New York: Doubleday, 1978.

Fabry, Charles, and Henri Buisson. "L'absorption de l'ultraviolet par l'ozone et la limite du spectre solaire." *Journal de physique théorique et appliquée* 3 (1913): 196–206.

Finkbeiner, Ann. *The Jasons: The Secret History of Science's Postwar Elite.* New York: Viking, 2006.

Finney, John W. "Pentagon Replies on Peril to Ozone." *New York Times*, Oct. 17, 1974.

Fleming, James Rodger. *Fixing the Sky: The Checkered History of Weather and Climate Control*. New York: Columbia University Press, 2010.

Foley, Henry M., and Malvin A. Ruderman. "Stratospheric NO Production from Past Nuclear Explosions." *Journal of Geophysical Research* 78 (1972): 4441–50.

———. "A Statement on Jason by Two Members." *Columbia Daily Spectator*, Apr. 28, 1972, 9.

Goldsmith, P., A. F. Tuck, J. S. Foot, E. L. Simmons, and R. L. Newson. "Nitrogen Oxides, Nuclear Weapon Testing, Concorde and Stratospheric Ozone." *Nature* 244 (1973): 545–51.

Haagen-Smit, A. J. "Atmospheric Reactions of Air Pollutants." *Industrial and Engineering Chemistry* 48 (1956): 65–70A.

———. "Chemistry and Physiology of Los Angeles Smog." *Industrial and Engineering Chemistry* 44 (1952): 1342–46.

Hampson, John. "Photochemical War on the Atmosphere." *Nature* 250 (1974): 189–91.

———. "Photolysis of Wet Ozone and Its Significance to Atmospheric Heating of the Ozone Layer." *International Council of the Aeronautical Sciences, Third Congress, Stockholm 1962*. London: Macmillan, 1964, 215–34.

Hartley, W. N. "On the Probable Absorption of the Solar Ray by Atmospheric Ozone." *Chemical News* (Nov. 26, 1880): 268.

Hunt, B. G. "The Need for a Modified Photochemical Theory of the Ozonosphere." *Journal of the Atmospheric Sciences* 23 (1966): 88–95.

Iklé, Fred C. "Foreword." *World-Wide Effects of Nuclear War*. Washington, DC: United States Arms Control and Disarmament Agency, 1975.

———. Letter to Philip Handler, Apr. 4, 1974. Reprinted in *Long-Term Worldwide Effects of Multiple Nuclear-Weapons Detonations*, by US National Research Council. Washington, DC: National Academy of Sciences, 1975.

———. "The Nether World of Nuclear Megatonnage." *Bulletin of the Atomic Scientists* 1 (1975): 20–25.

———. *Nuclear Disarmament without Secrecy*. Washington, DC: United States Arms Control and Disarmament Agency, 1974.

Johnston, Harold. "Reduction of Stratospheric Ozone by Nitrogen Oxide Catalysts from Supersonic Transport Exhaust." *Science* 173 (1971): 517–22.

Johnston, Harold, Gary Whitten, and John Birks. "Effect of Nuclear Explosions on Stratospheric Nitric Oxide and Ozone." *Journal of Geophysical Research* 78 (1973): 6107–35.

Lamb, Hubert H. "A Critical Problem." *Nature* 237 (1972): 53–54.

MacDonald, Gordon J. F. "How to Wreck the Environment." In *Unless Peace Comes:*

A Scientific Forecast of New Weapons, edited by Nigel Calder, 181–205. New York: Viking, 1968.

Miller, A. J., et al. "Nuclear Weapons Tests and Short-term Effects on Atmospheric Ozone." Paper presented at the Second International Conference on the Environmental Impact of Aerospace Operations in the High Atmosphere, American Institute of Aeronautics and Astronautics/American Meteorological Society, San Diego, CA, July 1974.

Mills, Michael J., et al. "Massive Global Ozone Loss Predicted Following Regional Nuclear Conflict." *Proceedings of the National Academy of Sciences* 14 (2008): 5307–12.

Molina, Mario J., and F. Sherwood Rowland. "Stratospheric Sink for Chlorofluoromethanes: Chlorine Atom-Catalysed Destruction of Ozone." *Nature* 249 (1974): 810–12.

Nelkin, Dorothy. *The University and Military Research: Moral Politics at M.I.T.* Ithaca, NY: Cornell University Press, 1972.

Nixon, Richard. "President's Message." *Environmental Quality, First Annual Report of the Council on Environmental Quality*. Washington, DC: US Government Printing Office, 1970, vii.

Sullivan, Walter. "Scientists Plan Series of Tests to Assess the Impact of Supersonic Air Traffic on Life and Climate." *New York Times*, Nov. 19, 1972.

US Congress. Office of Technology Assessment. *The Effects of Nuclear War*. Washington, DC: US Government Printing Office, 1979.

US National Research Council. *Long-Term Worldwide Effects of Multiple Nuclear-Weapons Detonations*. Washington, DC: National Academy of Sciences, 1975.

Whitten, R. C., W. J. Borucki, and R. P. Turco. "Possible Ozone Depletions Following Nuclear-Explosions." *Nature* 257 (1975): 38–39.

11

WHO OWNS THE AIR?

Contemporary Art Addresses the Climate Crisis

Andrea Polli

THE ACCELERATING CRISIS in climate change and the realization that humans are the primary cause of this change has raised questions about ownership and responsibility. Who "owns" the climate change crisis, and who is responsible for mitigating and reversing it if possible? The overwhelming response to these questions by governments internationally has been to propose a market solution, in essence, to sell the atmosphere. This chapter explores the idea of air for sale from economic, political, and cultural arts perspectives, and asks, "Can art impact the science and policy of climate change?"

Background

In 1644 Evangelista Torricelli, the Italian physicist, mathematician, and inventor of the barometer, wrote in a letter to Michelangelo Riccui, "We live submerged at the bottom of an ocean of air." In the words of Australian activist author Tim Flannery, the atmosphere is Earth's "energetic onion skin."[1] This thin, moving coating of the Earth is actually made up of five layers. The lowest layer is the troposphere, which, in middle latitudes, extends an average of eleven miles above the Earth and contains about 80 percent of all the atmosphere's gases, almost all of its water vapor, and most of its weather. An isothermal zone called the tropopause caps convection in this turbulently mixed layer. The next layer is called the stratosphere, so named for its stable stratification—containing an ozone-rich zone extending to a height of about thirty miles. The mesosphere, where the gases are extremely rarefied and the coolest temperatures of about $-90°C$ are found, extends to an altitude of about fifty-three miles. These three layers together are sometimes called the homosphere, since the gases here are typically well-mixed. The atmosphere changes again, drastically, in the next layer of very hot (up to 2,000°C), ionized gases, variously called the thermosphere (because it gets so hot), the ionosphere (because of its electromagnetic characteristics), and the heterosphere (because the gases are stratified according to their molecular weights). The base of the next layer, the exosphere, begins at 310 miles and extends to an altitude of about 620 miles where it merges with interplanetary space. In this layer the atmosphere ends, since it is too thin for the molecules to interact with one another as a gas. Most of this chapter will discuss naturally occurring and anthropogenically produced gases in the troposphere specifically, except in the parts referring to the ozone layer, which will concern the stratosphere.[2]

The air around us is composed of 78.08 percent nitrogen (N_2), 20.95 percent oxygen (O_2), 0.93 percent argon (Ar), and about 0.04 percent carbon dioxide (CO_2), with this latter constituent rising about 30 percent since the nineteenth century. Water vapor (H_2O) is a variable gas accounting for 1–4 percent of the atmosphere near the surface. The rest of the compounds are present in much smaller amounts—in descending order of abundance: neon (Ne), helium (He), methane (CH_4), krypton (Kr), hydrogen (H_2), nitrous oxide (N_2O), carbon monoxide (CO), xenon (Xe), ozone (O_3), and nitrogen dioxide (NO_2). Concentrations of any of the so-called trace gases can be higher due to local conditions. The concentration of the radioactive element radon (Rn), for example, can be 100 to 1,000 times the natural background concentration in enclosed spaces such as basements. These are the naturally occurring gases.[3]

The Clean Air Act of 1963 (public law 88–206) and its many revisions, notably that of 1967, which defined air quality control regions, was aimed at the prevention and abatement of air pollution. The Clean Air Act of 1970 (public law 91–604), a complete rewrite, "set National Ambient Air Quality Standards (NAAQS), to protect public health and welfare, and New Source Performance Standards (NSPS), that strictly regulated emissions of a new source entering an area. Standards were also set for hazardous emissions and emissions from motor vehicles. . . . Also, as a new principle, this Clean Air Act allowed citizens the right to take legal action against anyone or any organization, including the government, who is in violation of the emissions standards."[4] The Clean Air Act of 1990 (public law 101–549), "an Act to amend the Clean Air Act to provide for attainment and maintenance of health protective national ambient air quality standards, and for other purposes," designated states as being responsible for nonattainment areas, raised automobile emissions standards, incentivized the use of low-sulfur fuels as well as alternative fuels as a means of reducing sulfur dioxide—which is a main component of acid precipitation—in the atmosphere, mandated the installment of the best available control technology (BACT) to reduce the amount of air toxics, and called for a reduction in the amount of chlorofluorocarbons (CFCs) being used as a way of preventing ozone depletion. [5]

The Clean Air Act defines what are called "criteria air pollutants," or "emissions that cause or contribute to air pollution that may reasonably be anticipated to endanger public health or welfare," and whose "presence in the ambient air results from numerous or diverse mobile or stationary sources."[6] It authorizes the US Environmental Protection Agency (EPA) to enforce this law.[7] The criteria pollutants are carbon monoxide (CO), lead (Pb), nitrogen oxides (NO_x), ozone (O_3), particulate matter (PM), and sulfur dioxide (SO_2). There are also a large number of other compounds that have been determined to be hazardous called "air toxics." All told, in the United States, 188 gases are currently classified as pollutants.

IMPORTANT POLLUTANTS AND GREENHOUSE GASES

Greenhouse gases are good absorbers and emitters of infrared radiation that serve to warm the Earth. There are about thirty greenhouse gases; the primary ones are water vapor (H_2O), carbon dioxide (CO_2)—which is used as the yardstick for measuring the warming potential of other anthropogenic greenhouse gases, and methane (CH_4). Some greenhouse gases are classified as pollutants, but others, including CO_2, are not. Recently, the EPA has decided to classify rising carbon dioxide emissions as a hazard to human health.

Water vapor is the most powerful greenhouse gas, since it is much more abundant than all the others. It is not currently seen as a driver of anthropogenic climate change, serving primarily as a magnifier of other effects. Its saturation vapor pressure increases with increasing temperatures, and thus water vapor serves as a positive feedback mechanism for global warming. However, there are reports that the water vapor concentration of the atmosphere may be increasing due to human activities such as irrigation.[8] The fact that water vapor forms clouds further complicates its role in climate. By blocking sunlight, water vapor in the form of clouds can also have a cooling effect. High, thin clouds promote warming; low, thick clouds promote cooling. Of course all the fresh and salt water reservoirs of the blue planet, including lakes, rivers, oceans, and ice sheets are made of H_2O.

Carbon dioxide is a natural and essential component of the biosphere and anchors the natural carbon cycle. Human activity, such as burning fossil fuels, deforestation, and manufacturing cement, has caused its concentration to increase since about 1900. Of the CO_2 emitted by burning fossil fuels, 41 percent is from coal, 39 percent is from oil, and 20 percent is from natural gas. Because of the oxygen added during combustion, burning a ton of coal creates 3.67 tons of carbon dioxide. CO_2 is necessary for plant growth, although some plants grown experimentally in high-CO_2 environments have been found to have reduced nutritional value. Trees benefit much more from increased CO_2 than the grasses that are the staple of the human diet (rice, wheat, corn). Climate change may decrease crop yields because of shifting growth zones and moisture patterns, and other changes in ecosystems. There is fifty times more CO_2 in the ocean than in the atmosphere, but too much CO_2 in the ocean reduces the pH of seawater, placing stress on carbonate forms of life. There is presently around 400 ppm of CO_2 in the Earth's atmosphere, and this amount has been rising at about two ppm per year. The amount is very small, but it has a profound effect on warming. According to the Intergovernmental Panel on Climate Change (IPCC), attempts to stabilize CO_2 concentrations at 450 to 550 ppm would require a reduction in emissions of 60–80 percent below present levels.[9]

Methane accounts for about 10 percent of all US anthropogenic greenhouse gas emissions. In addition to natural sources such as wetlands, about 60 percent of the total emissions come from human activities such as leakage from natural gas systems and the raising of livestock. The concentration of methane has approximately doubled in the last two hundred years, but has leveled off in the twenty-first century. As a greenhouse gas it is sixty times more efficient in absorbing and emitting infrared radiation than CO_2, but being biologically active, has a much shorter residence time in the atmosphere. Scientists have assigned a global warming potential (GWP) to each gas based on its atmospheric concentration, residence time, and absorption cross-section, or ability to absorb energy. Note that all greenhouse gases are at least triatomic, allowing infrared energy absorbed to be expressed in modes of molecular vibration and rotation. The 100-year GWP of carbon dioxide is 1, while methane is rated 25, nitrous oxide 298, and some more esoteric gases such as HCFCs and sulfur hexafluorine are rated in the thousands to tens of thousands.[10]

Ozone is a tropospheric pollutant, a greenhouse gas, and in the stratosphere provides a layer of protection against harmful UV light.[11] The EPA regularly monitors surface ozone and provides public warnings on days with very high levels. Studies have found that ozone causes shortness of breath and coughing and triggers asthma attacks, and that high levels cause an increase in emergency room visits and hospital admissions. Ozone is an extremely reactive gas that irritates the respiratory system and can kill people with severe respiratory problems, and in fact studies show that as ozone levels increase, the risk of premature death increases. In the troposphere, ozone is caused by smog that forms when nitrogen oxide and hydrocarbons released into the atmosphere interact with one another, often because of sunlight, heat, and stagnant air. Ozone is one of the most dangerous components of smog, and in the past one hundred years there has been a global increase in ozone at ground level, particularly in urban areas.[12] However, in the stratosphere ozone occurs naturally and is necessary in shielding the surface from dangerous UV radiation.[13] Ozone depletion in the stratosphere contributes to higher rates of skin cancer, higher rates of cataracts, and other vision problems. A rhyme the EPA uses to describe this ozone distribution is "good up high, bad nearby."

The CFC and HCFC family of chemicals are powerful greenhouse gases that also destroy ozone in the stratosphere. Chlorofluorocarbons are compounds containing chlorine, fluorine, and carbon only; they contain no hydrogen. Hydrochlorofluorocarbons are a class of haloalkanes in which not all hydrogen has been replaced by chlorine or fluorine. These human-invented compounds were used widely as refrigerants, propellants, and cleaning solvents, but because of their deleterious effects on the ozone layer much of their use was prohibited by the Montreal Protocol and its amendments. HCFCs are used primarily as CFC substitutes, as they are more biodegradable and their ozone-depleting effects are only about 10 percent that of the CFCs.[14]

Sulfur dioxide is released when certain kinds of coal are burned and lasts a few weeks in the atmosphere before it is washed out by precipitation. It is not a greenhouse gas, and as an aerosol it actually has a cooling effect. However, it is a dangerous pollutant that causes acid rain by reacting with water to create sulfuric acid. Acid rain stresses plants and animals, makes lakes acidic, and damages tree growth, sometimes killing whole forests. In addition, SO_2 can affect human health, affecting people suffering from asthma and chronic lung diseases in particular.

Particulate matter (PM) includes both anthropogenic and naturally occurring solid particles and liquid droplets found in air such as smoke, turpenes, and other particles. They are identified by size (in microns) in pollution measurements as PM 2.5 or PM 10. PM typically lasts only a few weeks in the atmosphere and can have a cooling effect when it scatters light, making cloud particles more reflective. Alternatively, black carbon is now seen to be a major source of climate warming. Soot deposited on otherwise white snowbanks and ice sheets can also have a heating effect, since it absorbs sunlight. PM has also been shown to induce heart attacks and strokes, trigger asthma attacks, be implicated in lung cancer, and, in general, increase the need for medical care and hospital visits. People with cardiovascular diseases, children, and the elderly are most vulnerable to the health risks associated with PM, as are people who smoke heavily or suffer from chronic lung disease.

Ecologist Eugene F. Stoermer named the current era the "Anthropocene," or the age of humanity, when human activity has a significant global impact on Earth's ecosystems. Atmospheric scientist Paul Crutzen popularized this term. He estimates this age started at around 1800 CE, or at the beginning of the industrial age, because since then human population and human influence on the environment have risen dramatically. However, paleoclimatologist William F. Ruddiman thinks dramatic human influence on the environment dates to the beginning of agriculture, some 8,000 years ago.

How Is Air Quality Monitored?

Air quality is monitored for various reasons using a number of different methods, the most accurate being a combination of ground-based flask sampling and continuous monitoring, flask sampling at tall towers, and aircraft and satellite remote sensing. The National Oceanic and Atmospheric Administration (NOAA) operates an international system of flask sampling in which volunteers from sixty sites around the world travel to coastlines, hike up mountains, or walk miles into the desert once a week in order to fill two glass flasks with air using a battery-operated pump and compressor. After several weeks, these volunteers deliver the flasks to a mailroom in a US embassy, a nearby meteorological agency, or a university department, from which the flasks are returned to a laboratory in Boulder, Colorado. There, at NOAA's Earth Science Research Laboratory (ESRL), scientists in the Global Monitoring Division analyze the air to determine the global mix of greenhouse gases. "We get about 15,000 [flasks] a year," said Russell Schnell, director of observatory and global network operations in the Global Monitoring Division.[15]

On the local level, the state of California recently implemented a comprehensive air monitoring system. A bill passed by the state legislature in 2007 requires California to reduce its carbon emissions to 1990 levels by 2020, a reduction of 25 percent. By 2050, carbon emissions must be reduced 80 percent below 1990 levels. Given that California has the fifth-largest economy in the world, with the highest greenhouse gas emissions of any state, estimated at around 400 million metric tons per year, both these targets require substantial reductions. The monitoring system for the state of California, intended to help it achieve its emissions reductions goals, is called the California Greenhouse Gas Emissions Project (CALGEM); it is administered by Lawrence Berkeley National Laboratory.[16]

The Buying and Selling of "Emissions"

Andrew Simms, policy director of the New Economics Foundation (NEF) and the Foundation for the Economics of Sustainability (Feasta), has commented, "Economic super-powers have been as successful today in their disproportionate occupation of the atmosphere with carbon emissions as they were in their military occupation of the terrestrial world in colonial times."[17] In 1997, the Kyoto Protocol was established, requiring thirty-five industrialized countries and the European Union to reduce greenhouse gas emissions by an average of 5 percent below 1990 levels by 2012. Despite being the world's biggest emitter of greenhouse gases, the United States is not a signatory to the protocol. In the Kyoto Protocol, the main "basket" of greenhouse gases to be reduced includes carbon dioxide, methane, nitrous oxide, hydrofluorocarbons, perfluorocarbons, and sulfur hexafluoride. In addition, the protocol identifies "annex" gases (or indirect greenhouse gases) that developed countries must also monitor. These include carbon monoxide, nitrogen oxides, non-methane volatile organic compounds, and sulfur oxides.

The Kyoto Protocol includes a global greenhouse gas emissions trading system that has now been in place in Europe for over two years. This emissions trading system could also be called a "cap-and-trade" system, and this is how it works: in the first year, credits are generous, the total amount of emissions for each company is determined, and each company gets close to that amount. Then, each subsequent year, the amount of credits allotted to each company is reduced, allowing companies to lower emissions slowly. Where does the idea of addressing emissions through a market-driven economic solution come from? Consider some historical models:

The Chicago Climate Exchange (CCX) was a voluntary global emissions trading system launched in 2003 that covers six greenhouse gases. Participating companies committed to reduce their baseline greenhouse gas emissions by 6 percent between 2007 and 2010. The CCX, now defunct, traded over one million tons of CO_2 in its first six months of carbon trading, but as a voluntary system this was nowhere near the scale needed to make the necessary reductions. The troubled exchange became moribund in early 2010 and was closed in November with the carbon credit price per metric ton of CO_2 at $0.05, down from a high of $7.50 two years earlier.

In 1990, the United States launched a cap-and-trade program in sulfur dioxide as an amendment to the Clean Air Act. An initial objective of the program was to reduce sulfur dioxide emissions from utilities by 8.5 million tons below 1980 levels. To accomplish this, electric utility plants above a certain size were given an initial allocation of emissions allowances based on historical patterns. The Act included stiff penalties for excess emissions, at a value more than ten times that of reduction costs. The program achieved a high degree of compliance, and the current European greenhouse gas trading system is modeled after this system.[18]

Another model is the Montreal Protocol on Substances that Deplete the Ozone Layer, an international treaty that entered into force on January 1, 1989, designed to protect the ozone layer by phasing out the production of a number of substances believed to be responsible for ozone depletion, in particular CFCs and halons. The Montreal Protocol created a system for the international trading of allowances. In the protocol, trading is combined with a tax, to offset any large profits from allowances that might discourage the reduction of CFCs. Since the Montreal Protocol came into effect, the atmospheric concentrations of the most important CFCs and related chlorinated hydrocarbons have either leveled off or decreased, and Kofi Annan, former Secretary General of the United Nations, called the protocol "perhaps the single most successful international agreement to date."[19] However, by 1997 smuggling CFCs from developing nations, where they were still permitted, into the United States and other developed nations had become big business. In Miami in 1997 smuggling CFCs was believed to be second only to smuggling cocaine in cash value. CFC production effectively ended in 2005, but that year several companies in eastern China were found to be involved in illegal international trading of CFCs.[20] In 2007 a UN-sponsored summit agreed to accelerate the elimination of HCFCs by 2020, with developing nations following by 2030.

The difference between a CO_2 market and both the CFC and sulfur dioxide markets is that technology exists to clean up CFC and SO_2 emissions, and the problem applies only to a relatively small number of companies with outdated technology, while the CO_2 system applies to thousands of companies. Since coal is an abundant fuel, but quite dirty when burned, so-called clean coal technology is being promoted. These processes involve chemically washing the coal of its impurities, producing coal gas (to reduce carbon emissions), and scrubbing impurities from and capturing carbon dioxide in the stack gases. As yet, however, there are no techniques available to sequester carbon emissions cheaply or safely, making a market in CO_2 emissions reduction much less viable than the SO_2 market.

AIR AND PROPERTY RIGHTS

This brings us to a fundamental aspect of the greenhouse gas emissions trading system, the issue of property rights to the air. The idea of an environmental and natural resources economics came from the understanding that environmental resources are finite, and since these resources can be destroyed, there should be incentives for protecting them. Ecological economics provides both a mechanism for the valuation of environmental resources and an incentive for keeping within an established environmental "budget."[21] In 1997, the US Congress described it in this way:

> From an economic perspective, pollution problems are caused by a lack of clearly defined and enforced property rights. Smokestack emissions, for example, are deposited into the air because the air is often treated as a common good, available for all to use as they please, even as a disposal site. Not surprisingly, this apparently free good is overused. A primary and appropriate role for government in supporting the market economy is the definition and enforcement of property rights. Defining rights for use of the atmosphere, lakes, and rivers is critical to prevent their overuse. Once legal entitlement has been established, markets can be employed to exchange these rights as a means of improving economic efficiency.[22]

Emissions trading systems have long been criticized. Arguments for and against these systems range from concerns about flaws and possible abuses to fundamental critiques about ownership. The biggest beneficiaries of a greenhouse gas emissions trading scheme would be the banking industry (in the United States, this industry lobbied heavily for the implementation of a carbon cap-and-trade system) and the nuclear power industry, which stands to benefit from a loosening of restrictions on the production of new power plants.

However, despite the support of banks, not all economic experts back the use of such a system. Many would rather see a carbon tax in place. In practice, a carbon tax functions very much like a trading system: polluting companies either pay a tax or pay for carbon credits for their emissions. Both systems would also raise consumer prices on fossil fuels. Critics of the tax system say that it doesn't provide the incentive or "race for the pot of gold" that the carbon trading system provides by financially rewarding companies that can substantially cut emissions.[23]

Nicole Gelinas of the *Wall Street Journal* argues against a trading system in the United States from the perspective of global competition. She says that if US-based energy companies can't limit emissions, the international cap-and-trade system would allow them to buy emissions credits from other countries. She calls this a "direct subsidy to developing nations by paying for their power-plant upgrades." In other words, a federal carbon tax, an alternative to the international carbon cap-and-trade system, would provide revenue to the US government that could then be used to subsidize US power plant upgrades, while an international cap-and-trade system could put this revenue in the hands of other countries, particularly those that already have reduced emissions. Gelinas's views seem to be in line with the US government, as one of the reasons for why the United States has stated it is against the Kyoto Protocol is that it provides exceptions for developing countries.[24]

In *The Weather Makers*, Tim Flannery argues with this position, stating that although developing nations were not bound by the Montreal Protocol, it was very successful. In general, however, Flannery is mixed on the subject of carbon trading. He talks about the case of eastern European countries whose economies have suffered ruin since the 1990s and therefore are producing approximately 25 percent less CO_2 than they were in 1990. Since the Kyoto trading system requires emissions only 8 percent less than 1990 levels, these countries end up with a surplus of carbon credits, known as "hot air" to critics of the protocol.[25] Flannery also says that creating a new global currency is too risky, since the foundation of any currency is trust—in this case trust that the seller will lower emissions—and he doesn't see any guarantee that this will happen. However, he observes that emissions trading is cost-effective, and as a tool to reduce pollution it has been successful in the past, using the example of sulfur dioxide trading.

In *Carbon Trading*, Larry Lohman quotes Flannery's work when outlining the potential effects of climate change, but is much more pessimistic than Flannery with regard to markets. He looks at the seriousness of the problem and predicts that many markets will collapse. He gives the example of the insurance business, providing quotes from insurance specialists who estimate that chaotic climate change could cause insurance rates to increase to several times that of world economic growth, creating a situation where the world economy will not be able to sustain the losses and will collapse. Lohman is critical of the hope held by many industrialized countries that technological developments, such as carbon sequestration, will allow continued use of fossil fuels. He devotes a chapter to analyzing how these technological solutions are nothing but a smokescreen to distract public attention from the government's lack of necessary action. He sees this as a "second strategy," the first being denial of the existence of anthropogenic climate change: "The first strategy works to reshape or suppress understanding of the climate problem so that public reaction to it will present less of a political threat to corporations. The second strategy appeals to technological fixes as a way of bypassing the debate over fossil fuels while helping to spur innovations that can serve as new sources of profit. The third strategy appeals to a "market fix" that secures that property rights of heavy northern fossil fuel users over the world's carbon-absorbing capacity while creating new opportunities for corporate profit through trade."[26] In *Earth in the Balance*, Al Gore embraced in part this "second strategy," proposing a Strategic Environment Initiative (SEI), which, like the US Strategic Defense Initiative, would be a major national effort but with a focus on the environment rather than the military. Although he emphasizes that any new technologies must first be evaluated, the focus of his proposed SEI is on developing new technologies to combat climate change. Finally, Lohman is very critical of what he calls the third strategy, the "market fix," primarily because of the property rights issue, the privatization of the air. Like Flannery, he is concerned that the "top down" creation of a greenhouse gas emissions market, created without public debate, will create a market filled with distrust, a lack of faith in both the structure and the implementation of the market. The bottom line according to Lohman is that the current emissions trading structure gives the largest allowances to the biggest emitters, effectively handing out billions of dollars to the most egregious polluters and providing incentives to them to keep polluting.

In *The Great Emissions Rights Give-away*, Andrew Simms proposes an alternative structure for EU carbon trading.[27] In this structure, the EU's emissions allowances would be divided up on an equal per capita basis and distributed to every EU resident. Residents would then be able to sell these allowances to companies or keep them off the market, promoting cleaner air. Simms compares this approach to some other alternative approaches. Dr. David Fleming's tradable energy quotas proposal envisions governments providing each citizen with a portion of carbon units equal to the amount of the general public's fossil fuel use. Citizens then can use the units to purchase energy and fuel, or sell unused units on an open market. Carbon units used by industry are sold at an auction similar to the Federal Communications Commission broadcast rights auction. The Sky Trust in the United States proposes a system whereby any money made from the sale of emissions credits are kept in a trust to ensure that any financial gains are used to benefit the public. The proposals of both the Sky Trust and Simms's NEF/ Feasta are based on the fundamental principle that the atmosphere belongs to all people equally and not to governments or corporations.

SELLING AIR TO THE PUBLIC

One might think that the idea of "air for sale" is only an abstraction.[28] There are, however, many ways that air has been commercialized—for example, in the use of bottled oxygen in medicine and sports. Recreational uses include the rising popularity of something called the "oxygen bar" and canned air, where oxygen is touted as a cleansing and medical "therapy." Advertisements focus on the healing power of air, using aromatherapy and something marketers call "oxygen therapy." Advertisers promote the air as energizing for exercise and effective in combating cigarette smoke and curing a hangover. Their product is marketed on the idea of being pure, fresh, and clean; many promote it as an escape from the smog of city life. In the case of the oxygen bar, customers pay for a five-minute session or so, in which they are able to relax and breathe clean, sometimes scented, air. The oxygen bar started as a trend in the 1990s in Japan, Mexico, and South America and quickly spread to nightclubs, spas, casinos, and malls in Europe and the United States. In 2003, the oxygen bar at Olio!, a restaurant at the MGM Grand Hotel in Las Vegas, boasted 200 to 400 customers per day.[29] Portable canned air is becoming just as popular and widespread. In Japan, a recent large-scale commercial venture is O_2 supli, a portable can of oxygen. The oxygen comes in two flavors, "strong mint" (called the brain can) and "grapefruit" (called the body can) at a price of 600 yen ($7.50) a can: "The idea behind the product is to allow buyers to replenish their oxygen levels any time they feel a lack of it due to stress, fatigue, or other factors . . . Each can contains enough oxygen for 35 two-second inhalations, meaning each can lasts for roughly a week if it is used five or six times a day. At first the canned oxygen will be sold in Tokyo . . . then at all 11,000 of

Seven-Eleven Japan's nationwide stores."[30] It seems that we buy and sell all the elements: land, water, fire (fuel)—and now air.

THE SALE OF AIR IN CONTEMPORARY ART

According to art scholars Alexander Alberro and Sabeth Buchmann, "When art becomes idea, idea becomes commodity."[31] Perhaps the arts, specifically contemporary conceptual artworks, have played a role in making buying air culturally acceptable. As creative works, art and architecture have value in society—not just cultural value (although they have that too), but monetary value. In the 1950s and '60s, Yves Klein's idea of *Air Architecture* challenged the definitions of art and architecture, but on a wider scale may have unintentionally contributed to the idea of commodifying the public resource of air. Klein was interested in the ways that humans can use science and technology to conquer the ephemeral, to the point of turning even air and fire into building materials. Klein saw science and technology as the savior of architecture, promoting new forms and structures made from sculpting the air and other "immaterial-materials." He believed that *Air Architecture* would actually improve the environment, saying that "Air Architecture must be adapted to the natural conditions and situations, to the mountains, valleys, monsoons, etc., if possible without requiring the use of great artificial modifications."[32]

In the late 1960s a group of artists including Robert Barry became associated with dealer Seth Siegelaub and started producing work that questioned the limits of art. Barry's work, known as "invisible" art, included *The Inert Gas Series* (1969) in which a specific amount of gases such as neon, xenon and helium are released "from measured volume to indefinite expansion" in the Mojave Desert.[33] The critic Lucy Lippard observed in *Six Years: The Dematerialization of the Art Object from 1966 to 1972* that "novelty is the fuel of the art market," and at the time of Barry's *Inert Gas Series*, this "fuel" was being burned at a rapid pace, constantly stretching the boundaries of the definition of art. This "dematerialization" attempts to remove art from its status as commodity by creating such "objects" of art, like the natural expansion of gas, that would be absurd to commodify, yet this work doesn't remove itself from the art market. In fact, the dematerialization movement was being driven in large part by the art market looking for newer and more avant-garde ideas.[34] These ties to the market, necessary for the creation of the art and the survival of the artist, nonetheless create a paradoxical situation in which the immaterial is moved into the object realm. The critical stance of the artist on the art market is completely lost through the positioning of the work within the art market.

Tue Greenfort's *Bonaqua Condensation Cube* of 2005 pays homage to Hans Haacke's *Condensation Cube* of 1963. The contemporary work uses Bonaqua, popular brand of bottled water, as the water of condensation. Greenfort is directly addressing the issue of ownership. What was considered in 1963 to be a public resource by 2005 was a commercial product. Like the earlier work, the piece is positioned in a gallery as an object with at least the expectation of being given a monetary value. Also like the earlier work, this piece pokes fun at the absurdity of the commercial-gallery system, but problematically both remain a part of that system.

Here are some examples of recent works related to air that don't reside in the gallery context but rather in the context of public art. Laurie Palmer's 2005 *Hays Woods/Oxygen Bar* project at Carnegie Mellon University highlights the natural processes that create air, and highlights air as a public resource:

> Want to breathe some pure oxygen courtesy of plants from Hays Woods?
> Try the Oxygen Bar, a project of artist A. Laurie Palmer. The oxygen bar is a mobile breathing machine, offering free hits of "natural" oxygen on a first-come, first-serve basis. This oxygen is produced by the photosynthetic work of green plants (from Hays Woods) and is offered as a public service. It reproduces in miniature the beneficial cleansing and refreshing effects of city green spaces on the air we breathe. The oxygen bar anticipates the imminent loss of public resources that filter Pittsburgh's dirty air and replenish it with oxygen—in particular, Hays Woods. At the same time, the oxygen bar anticipates the active participation of citizens of Allegheny County in land use decisions affecting our public health.[35]

The premise of Amy Balkin's 2007 work *Public Smog* lies in the economic system of emissions trading. The "global public" purchases as many emissions allowances as possible on the emissions market. These carbon offsets are then retired—in other words, taken off the market, making them unavailable to polluting corporations. By openly embracing the free market for the public good, Balkin presents a sharp critique of the system the project must operate within. The "public space" in which her work operates is the emissions trading market.

Balkin's work clearly questions the emissions trading system. I believe that the solution she proposes is meant to be absurd. But, in the context of contemporary culture, the solution seems like a viable one—in fact, it's very similar to the structure proposed by Feasta, in which people buy and sell emissions credits on the market, except in the case of Feasta, a certain amount of emissions credits are distributed for free to citizens. A group called TheCompensators* proposes the exact same solution as Balkin does, claiming to have retired over 1,500 EU emissions allowances.[36] These solutions create potential problems grounded in the very idea of the market. Healthy markets grow, and if people decide to buy emissions credits or clean air, what could happen to the market is that more emissions credits will be issued to balance the market and meet demand, in effect forcing the public to pay ever-higher prices for clean air.

The difference between Balkin's *Public Smog* and TheCompensators* projects lies not only in that one identifies itself as an art project and the other as an environmental project, but in the underlying metaphors that support the work. By using the metaphor of a public park, Balkin's work allows viewers and participants to look at the system of emissions trading through a familiar lens. Most viewers, understanding the difference between public and private property in the context of land and the implications of privatizing public land, gain a greater understanding of the implications of an emissions trading system through this metaphor. The message of *Public Smog* becomes even clearer when seen in relation to one of Balkin's earlier works, *This Is the Public Domain*. In 2003, Balkin purchased a 2.5-acre parcel of land in Tehachapi, California, with the intent of creating a permanent, international commons, free to everyone, held in perpetuity. Since there is no precedent to creating such a commons in the current legal system, Balkin's project involves a complicated legal process that explores solutions in both real property law and copyright law. Through this process, Balkin questions the foundations of the existing laws. The classification of *This Is the Public Domain* as an art project is essential to Balkin's legal approach, which explores the land as a creative work of art where there is a movement of people who move fluidly between the roles of artist, environmentalist, and activist. Ben Engebreth's work *Personal Kyoto* provides individuals in various US cities the chance to comply personally with the Kyoto Protocol by monitoring their energy use with the goal of reducing their personal emission of greenhouse gases. Engebreth himself doesn't identify purely as an artist or call this project an art project, and the project operates outside of the art market on the internet. *Personal Kyoto* and other projects that focus on personal responsibility embrace the philosophy of "voluntary" emissions reduction, like the (now closed) Chicago Climate Exchange on an individual scale. Does this focus on personal and corporate choice and responsibility detract from the urgent need to curb emissions now? Is the American individualist ideal, the ideals behind the problematic proposal recently made by the United States to the EU promoting volunteer emissions reduction, going to promote the changes needed? Voluntary emissions reductions may be a great idea on an individual level, but will they work on a corporate scale?

The questions raised by the works discussed here do not represent a criticism of the artworks; the artists should be praised for raising these complex questions. The paradoxical problems are a function of the systems in which the works exist, either the gallery art world, with an economy based on the buying and selling of works, or the public art world, in which works are owned by government or private interests, including those works which operate in the semi-public forums of the market or internet. In the context of climate change, the works bring up larger questions about the potential of art in a time of global environmental crisis, and more specifically the potential of art and science collaboration.

The *Airlight* Project

Airlight is the name given to a visible white smog caused by the illumination of fine dust particles in the air. The term is often used in Los Angeles, where fumes from car exhaust create airlight, what author Lawrence Weschler describes as "a billion tiny suns."[37] My *Airlight* series first began as *Airlight Taipei* in summer 2006. Summer in Taipei is unbearably hot and humid, forcing residents to stay in air-conditioned buildings most of the day. The city is crowded, with over six million in the greater Taipei area. Although public transportation is excellent, several elevated highways cut through the city, like contrails cutting through the dense air. Taipei's geography works against its air quality. Taipei is located at the base of a bowl, surrounded on all sides by small mountains with only one small outlet for the stagnant air that often stays trapped for days. In addition, Taipei is downwind of southern China, where the energy demands of recent modernization have meant the development of more coal-burning power plants. Windflow from west to east puts a large amount of the pollution from China's coal industry into the air in Taipei. The effects of poor air quality are visible in the faces of its citizens, or rather, visible over the faces of citizens, as dust masks have become a fashion item, color-coordinated with clothing and motorbike helmets.

During a residency at the Taipei Artist Village, I had the great fortune to meet and collaborate with Dr. Chung-Ming Liu, director of the Global

Change Research Center and professor in the Department of Atmospheric Sciences at National Taiwan University. For our project, Dr. Liu gathered and formatted real-time Taipei air quality data for almost twenty sites around the city onto a website. This allowed me to automatically download hourly amounts of particle pollution, ozone, and other pollutants in the atmosphere and translate this information in real-time into a changing rhythmic visual and soundscape, translating the "noise" of the pollutants into a kind of rhythmic "noise" that expressed what Dr. Liu called the "daily variation" of air quality in the city. The traffic engineering office of Taipei city offers a large number of public traffic cameras, so I was able to synchronize the sound of the air quality with live traffic webcam images. I used the pollutant levels to make the images break apart, appearing and disappearing with rising and falling pollutant levels.

The idea of "noise" was central to the structure of both the sound and image in this work. I used a source sound that had a wide frequency spectrum—in other words, a very "noisy" sound. I then used the levels of pollutants to amplify and filter frequencies in this noise, picking out certain frequencies to represent levels of pollutants, and creating the effect of high-pitched screeches, like automobile brakes, when pollutant levels were high. Despite this noisy structure, the resulting sound would loop, with the loop changing only slightly each hour with a new reading. This repetitive structure created a rhythmic, ambient sound that functioned very much like a background soundscape. The imagery was also structured around the idea of noise. The original image was an unaltered traffic cam image that would pixelate based on the levels of pollutants in the air. This has the effect of a blurring and focusing of the image, in a rhythmical way in time with the sound. The rhythmic blurring and focusing of the image gave the effect of quivering or breathing, giving the image a kind of life. In discussing ephemeral and process-based art, Steven Connor says that "in much recent art, air has become the marker, not of the difference between art and life, but of the aspiration of art to trespass beyond its assigned precincts, to approach and merge into the condition of "life."[38] In the *Airlight* series, I have attempted to give a kind of "life" to the air quality data being collected, at times creating an alarming scream and image blur that increases in intensity as the levels of pollutants increase.[39]

Since *Airlight Taipei* was first shown, I have been able to create similar projects in two other locations: Southern California and Boulder, Colorado. The first project was *Airlight Socal*, for which I was able to work with live webcams from the California Department of Transportation (DOT) and daily amounts of O$_3$ and NO$_2$ for various locations in Southern California updated hourly provided by the South Coast Air Quality Management District with the help of Kevin Durkee. For *Airlight Boulder*, I integrated hourly air quality data provided by the AQI and VSI Air Quality Reporting Systems of the Colorado Department of Public Health and Environment and webcam images from the Colorado DOT.

Air quality data is most often presented to the public through preformatted webpages that include charts, graphs, and color-coded alerts. For the *Airlight* series, it is necessary for providers to allow access to the raw data, and each organization I worked with was able to provide this data to me on an individual basis. Through the Boulder project, I came in contact with the AIRNow Data Management Center, which collects and distributes air quality data from over 2,000 sites in the United States. They were very interested in the work I was doing, and are interested in ways in which the data being collected can be communicated to the public more effectively. I communicated extensively with AIRNow staff about the kind of data formatting I need for the project, and they used the *Airlight* project as a model for implementing a way to provide raw data access to artists and developers around the world through the web interface. Presently, developers need to contact AIRNow for the specific URLs to access the raw data, but hopefully they will create a more open platform in the future as the public realizes the importance of air quality data.

To me, this essay raises more questions than answers. Weather and climate research involves unavoidable uncertainties. These uncertainties have been used to discredit science and have been exploited to support various political agendas. Art often presents a personal interpretation of information. How does the personal expression element of art intersect with the uncertainties of weather and climate science? Can this intersection help or hinder public understanding of science? What is the responsibility of the artist to present "correct" science in art/science collaborations? With an element of uncertainty, how is "correct" science to be presented? Currently in the United States science is highly politicized, particularly the science of climate change. How do these politics affect art and science collaborations dealing with climate change? Can art help science out of the political quagmire?

Because artworks operate within various economies, whether they are gallery works or works of public art, how does this positioning in the marketplace affect the message and effectiveness of the work? In particular, how does the marketplace affect the interpretation of works addressing the urgent issue of climate change? Must artworks always be seen in the context of some market? Is this beneficial to addressing issues related to air quality and climate change? What other mechanisms exist or can be created for art and art/science collaborations so that artwork can be experienced without a market bias? Structures within academia for artistic production and scientific research and within independent community- and publicly funded organizations internationally can serve to promote and support art and science collaborations and science communication outside the market system. The benefits of such structures are not only to encourage a critical dialogue about existing initiatives but to explore other models of exchange in order to improve our environment for generations to come.

NOTES

1. Flannery, *Weather Makers*, 5.

2. Anthes, *Meteorology*, ch. 1.

3. Ibid.

4. James Rodger Fleming and Bethany Knorr, "History of the Clean Air Act: A Guide to Clean Air Legislation Past and Present," 1991, http://www.ametsoc.org/sloan/cleanair/.

5. Ibid.

6. United States Code, Title 42, 2010 edition, http://www.gpo.gov/fdsys/pkg/USCODE-2010-title42/html/USCODE-2010-title42-chap85-subchapI-partA-sec7408.htm.

7. For more on the EPA see Lee, this volume.

8. Oltmans et al., "Increase in Stratospheric Water Vapor."

9. IPCC, AR4.

10. IPCC, AR4, 212.

11. See Dörries, this volume.

12. See Brimblecombe, this volume.

13. See Dörries, this volume.

14. EPA, Oct. 11, 1997, http://www.epa.gov/fedrgstr/EPA-AIR/1995/October/Day-11/pr-1117.html.

15. USINFO interview, Aug. 24, 2007, http://usinfo.state.gov/xarchives/display.html?p=washfile-english&y=2007&m=August&x=20070829170618icnirellep4.142398e-02.

16. Science @ Berkeley Lab, Apr. 2007, http://www.lbl.gov/Science-Articles/Archive/sabl/2007/Apr/02-CALGEM.html.

17. Cited in Lohman, *Carbon Trading*, 19.

18. US Congress, Joint Economic Committee Study, July 1997, http://www.house.gov/jec/cost-gov/regs/cost/emission.htm.

19. "Stratospheric Ozone Depletion 10 Years after Montreal," naturalSCIENCE, Oct. 8, 1997, http://naturalscience.com/ns/cover/cover4.html.

20. "Corruption Stalls Government Attempts to Curb CFC Trade," *China Development Brief*, Dec. 15, 2005, http://www.chinadevelopmentbrief.com/node/371.

21. Tietenberg and Lewis, *Environmental and Natural Resources Economics*.

22. US Congress, Joint Economic Committee Study.

23. Deborah Solomon, "Climate Change's Great Divide," *Wall Street Journal*, Sept. 12, 2007, http://online.wsj.com/article/SB118955082446224332.html.

24. Nicole Gelinas, "A Carbon Tax Would Be Cleaner," *Wall Street Journal*, Aug. 23, 2007, http://www.manhattan-institute.org/html/_wsj-a_carbon_tax_would_be_cleaner.htm.

25. Flannery, *Weather Makers*, 225.

26. Lohman, *Carbon Trading*, 54.

27. "The Great Emissions Rights Giveaway," *New Economic Forum*, Feasta: Foundation for the Economics of Sustainability, Mar. 2006, http://www.feasta.org/documents/energy/emissions2006.pdf.

28. See George England, *The Air Trust* (1915), discussed in Fleming, *Fixing the Sky*, 36–38.

29. Jean Rimbach, "Oxygen Bar in Atlantic City, N.J. Feeds Fresh Air to Tourists," *Free Republic*, Dec. 31 2003, http://www.freerepublic.com/focus/f-news/845765/posts.

30. *Mainichi Daily News*, May 14, 2006.

31. Alberro and Buchmann, *Art after Conceptual Art*.

32. Klein, Noever, and Perrin, *Air Architecture*.

33. Wood, *Conceptual Art*, 35–36.

34. Lippard and Chandler, "Dematerialization," 31.

35. Palmer, "Hays Woods/Oxygen Bar."

36. TheCompensators*, May 24, 2012, http://www.thecompensators.org.

37. Wechsler, *Vermeer in Bosnia*.

38. Steven Connor, "Next to Nothing: The Arts of Air," talk given at Art Basel, 2007.

39. See http://www.andreapolli.com/airlight.

REFERENCES

Alberro, Alexander, and Sabeth Buchmann, eds. *Art after Conceptual Art*. Cambridge, MA: MIT Press, 2006.

Anthes, Richard. *Meteorology*, 7th ed. Upper Saddle River, NJ: Prentice Hall, 1997.

Eilperin, Juliet. "Bush Aims for Market Approach to Fishing." *Washington Post*, Sept. 20, 2005, A21.

Flannery, Tim. *The Weather Makers*. New York: Grove Press, 2005.

Fleming, James Rodger. *Fixing the Sky: The Checkered History of Weather and Climate Control*. New York: Columbia University Press, 2010.

Gore, Al. *Earth in the Balance: Ecology and the Human Spirit*. Boston: Houghton Mifflin, 1992.

Klein, Yves, Peter Noever, and Francois Perrin. *Air Architecture*. Los Angeles: MAK Center for Art and Architecture, 2004.

Lippard, Lucy. *Six Years: The Dematerialization of the Art Object from 1966 to 1972*. New York: Praeger, 1973.

Lippard, Lucy, and John Chandler. "The Dematerialization of Art." *Art International*, Feb. 1968, 31.

Lohman, Larry. *Carbon Trading: A Critical Conversation on Climate Change, Privatisation and Power*. Development Dialogue 48. Uppsala: Dag Hammarskjöld Foundation, 2006.

Oltmans, S. J., H. Voemel, D. J. Hofmann, K. H. Rosenlof, and D. Kley. "The Increase in Stratospheric Water Vapor from Balloon-Borne Frostpoint Hygrometer Measurements at Washington, D. C., and Boulder, Colorado." *Geophysical Research Letters* 27 (2000): 3453–56.

Palmer, Laurie. "Hays Woods/Oxygen Bar." Project proposal materials. Pittsburgh: Carnegie Mellon University, 2005.

Tietenberg, Tom, and Lynne Lewis. *Environmental and Natural Resources Economics*, 9th ed., Prentice Hall, 2011.

Wechsler, Laurence. *Vermeer in Bosnia*. New York: Pantheon, 2004.

Wood, Paul. *Conceptual Art*. Movements in Modern Art. London: Tate Publishing, 2002.

12

CARBON "DIE"-OXIDE
THE PERSONAL AND THE PLANETARY

James Rodger Fleming

CARBON DIOXIDE IS currently one of the most feared molecules on the planet. In trace amounts it is increasingly being linked to climate change. In much more concentrated amounts it is a narcotic or asphyxiating gas, known in antiquity as *spiritus letalis* and as "mephitic air" by early chemists; it has been employed, monitored, and controlled in modern times by anesthetists, brewers, butchers, divers, submariners, and astronauts, to mention but a few. Inhaled and exhaled in every breath, generated by human activity, and emitted in geologic fissures, carbon "die"-oxide has filled the dark crevasses and death valleys of the world, killing the unsuspecting. It may also be a "climate killer," toxic to civilization.[1] This is its story.

A journey across carbon dioxide mindscapes and landscapes includes exotic ideas and extreme places: *spiritus letalis*, the wild spirit of the forest, and fixed air; volcanic rims, valleys of death, dark and dank caverns; the global atmosphere, the global ocean, and the globe itself. As a toxic imaginary carbon dioxide resides in the ancient caverns of hell, in the fear and fascination of asphyxiation, and in the worldwide recent angst attending global climate warming. Travelers described poisoned landscapes, valleys and ravines where carbon dioxide lurked as a silent killer. The elite of Europe on the Grand Tour frequented the Grotto del Cane in Italy; intrepid explorers visited Batur, Java, reporting a valley of death; and early American naturalists described Death Gulch, an area in Yellowstone National Park devoid of trees and life. Danger, unpredictability, and astonishment surround these deathscapes. As understanding and awareness of the toxic effect of miniscule amounts of carbon dioxide on climate developed, so too did apprehensions grow of a looming global calamity.

In 1922 Willis Luther Moore, retired chief of the US Weather Bureau, wrote of Earth's "fourth atmosphere," carbon dioxide, coexisting with nitrogen, oxygen, and water vapor. "It forms the chief food supply of all green-leaved plants. It is as necessary to the life of vegetation as is oxygen in the supporting of animal life." Yet in sufficient concentrations it is a "poison," a gas that robs humans and animals of health and even life itself. Fifty percent denser than air, carbon dioxide is generated by combustion and respiration and has a tendency to collect in low places—cracks, crevices, sewers, cellars, breweries, sleeping rooms—unless forcefully expelled by larger atmospheric circulations.[2]

Carbon dioxide is omnipresent. It holds the balance of life and death, at once both toxic and restorative. It is plant food, and, as transformed by the biosphere, it is our food. The air we inhale contains about 0.04 percent carbon dioxide—an amount thought to be harmful to Earth's climate system. The air we exhale contains about 4 percent carbon dioxide—sufficient to cause discomfort and trigger headaches if rebreathed.[3] A thirty-minute exposure to 5 percent concentrations mimics intoxication, and a few minutes of breathing 7 to 10 percent concentrations will suffocate the average person, rendering him or her unconscious and on death's door.

In the nineteenth century the work of prominent researchers such as John Tyndall and Svante Arrhenius implicated various gases, including carbon dioxide, as possible agents of geological climate change (see below), but major scientific doubts led to an eclipse of this theory.[4] Until quite recently, carbon dioxide was not considered a climate-changing gas. For example, in 1936 the American Chemical Society published *Carbon Dioxide*, a monograph "aimed at a very broad treatment of the many phenomena in nature and industry in which carbon dioxide is an important factor."[5] The book contains no discussion at all of climate. This is quite surprising to modern readers steeped in the lore of carbon dioxide, a substance that has become the public's "environmental molecule of choice" and concern, implicated as the key agent in global climate change.[6]

This chapter is divided into three sections: the first on legendary and historical deadly caves and toxic valleys, next on the history of understanding and manipulating the substance that eventually came to be known as CO_2, and a final section that presents two possible meanings of carbon dioxide as a climate killer.

DEADLY CAVES AND TOXIC VALLEYS

There are numerous accounts of deadly caves and toxic valleys in classical and historical sources. According to legend, the Oracle at Delphi at the site of Apollo's birth on the slopes of Mount Parnassus inhaled fumes issuing forth from a crack in the ground and, intoxicated by the prophetic breeze (*manitkon pneuma*) that arose from the cleft in the rocks, allowed the deity to possess her spirit as she proffered visions of the future.[7] Plutarch, who once served as a priest at Delphi, told of a shepherd named Coretas who discovered the site when he fell into the crevasse there by accident and later gave forth inspired utterances and foretold the future.[8] Of the sentences written in Apollo's temple at Delphi, perhaps the most famous is the aphorism, "Know thyself."[9]

Some forty kilometers east of Delphi, the Cave of Trophonius served as a shrine to one of the legendary builders of the Temple of Apollo. Seeking enlightenment there was quite an ordeal, however, for "to descend into the cave of Trophonius" became an idiomatic expression meaning "to suffer a great fright." A seeker first underwent rituals of purification, lodged at the residence there for several days, bathed in the river Hercyna, and was anointed with oil and blessed by the priests. Numerous animal sacrifices were necessary—to Apollo, Chronos, Zeus, Hera, Demeter, and others—with a final ram offered to Trophonius. A diviner read the portents as revealed in the entrails. The petitioner then drank from the fountain of forgetfulness to clear his mind and from the fountain of memory in order to be able to recall his descent into the cave. After final oblations, the supplicant proceeded to the cave clad in a linen tunic girded with ribbons and, lying on his back holding ceremonial barley cakes kneaded with honey, descended feet-first into the chasm through a narrow masonry portal. After a sojourn there, where visions and revelations were *de rigeur*, the now-enlightened seeker of truth returned to the surface, again feet-first, where the priests interrogated him in the chair of memory and inscribed a tablet recounting his experiences. Finally, his relatives carried him, "paralyzed with terror and unconscious both of himself and of his surroundings," to the residence where, gradually, he re-

FIG. 12.1. Depiction of the Grotto del Cane. SOURCE: Geoffrey Martin, *Triumphs and Wonders of Modern Chemistry* (New York: Van Nostrand, 1913), 238.

covered his faculties.[10] Scholars have assumed that this state of intoxication was due to carbon dioxide and cited earlier accounts of delirium, cramps, dizziness, and ringing in the ears from such exposures.[11]

Pliny the Elder mentioned the exhalation of lethal vapor, *spiritus letalis*, from vents in the earth, notably the Temple of Mephitis, where all who enter die (except the elect).[12] He described sites near Pompeii that "exhale deadly vapors, either from caverns, or from certain unhealthy districts," calling them "Charon's sewers." Pliny also mentioned that marble, when pulverized and mixed with vinegar, was used by ancient physicians and surgeons as a painkilling liniment. Such a concoction would release carbon dioxide.[13] Virgil wrote of a fearful toxic cave:

> There's a place in Italy, at the foot of high mountains,
> famous, and mentioned by tradition, in many lands,
> the valley of Amsanctus: woods thick with leaves hem it in,
> darkly, on both sides, and in the center a roaring torrent
> makes the rocks echo, and coils in whirlpools.
> There a fearful cavern, a breathing-hole for cruel Dis,
> is shown, and a vast abyss, out of which Acheron bursts,
> holds open its baleful jaws, into which the Fury,
> that hated goddess, plunged, freeing earth and sky.[14]

Aeneas, seeking the golden bough, consulted the sibyl at "the foul jaws of stinking Avernus" and together they entered the mouth of a cave belching mephitic vapors that led to the realm of Dis and the river Acheron, where the grisly daemon Charon waited with his ferryboat.

> There was a deep stony cave, huge and gaping wide,
> sheltered by a dark lake and shadowy woods,
> over which nothing could extend its wings in safe flight,
> since such a breath flowed from those black jaws,
> and was carried to the overarching sky, that the Greeks
> called it by the name Aornos, that is Avernus, or the Bird-less.[15]

Such was the stuff of legends that enticed later travelers to seek out deadly caves and death valleys.

The Grotto del Cane (see fig. 12.1) on the shore of Lake Agnano near Naples was a popular site for travelers on the Grand Tour. It was named for the practice of sending dogs into a small cave where carbonic acid vapors pooled. For a small fee visitors could watch the keeper of the grotto hold a dog near the cave floor where it would swoon in a few moments. Then the attendant would throw the hapless creature into the lake to revive it.

> Behold, the cave murderer:
> An Ancient Rite.
> The Greyhound trusts you,
> But the heavy aer strikes oppressive,
> And stunned, the dog falls instantly.
> But surviving the peril of the river,
> Revives, and scrambling up its steep banks,
> Takes to the open fields,
> Its life force restored.[16]

Author, journalist, and philanthropist Grace Greenwood (1823–1904) wrote of visiting the cave during her grand tour in 1853. The little puppy of the day "came up very reluctantly to the trial. . . . Laid upon his back in the cave, he set up a low, piteous howl, which was answered by a sympathetic whine from several dogs without." Also being sympathetic, her party intervened before the expected onset of "strong convulsions." Yet, she wrote, "there was quite a satisfactory amount of spasmodic action, and the poor little brute tumbled about very clumsily, and walked drunk, up to the time of our leaving." She admitted it was a "cruel, senseless exhibition; but it is one of the sights which every body *does*, and, had we omitted it, certain it is that some kind of friend would afterwards have condoled with us on having missed 'the very best thing to be seen at Naples.'"[17]

Alexandre Dumas told of two unfortunate dogs, Castor and Pollux, "whose lives were intended to flow in perpetual fainting fits," and of the ragged urchins who gathered there seeking coins from English tourists, with a frog, a snake, a guinea pig, and a cat for hire. Legend has it that a Turkish viceroy brought two slaves here to die, "gratis." Johann Wolfgang von Goethe mentions the cave in his *Italian Voyage*. Mark Twain yearned, with morbid humor, to kill a dog in "the poisoned grotto":

> Everyone has written about the Grotto del Cane and its poisonous vapors, from Pliny down to Smith, and every tourist has held a dog over its floor by the legs to test the capabilities of the place. The dog dies in a minute and a half—a chicken instantly. As a general thing, strangers who crawl in there to sleep do not get up until they are called. And then they don't either. The stranger that ventures to sleep there takes a permanent contract. I longed to see this grotto. I resolved to take a dog and hold him myself; suffocate him a little, and time him; suffocate him some more and then finish him. We reached the grotto at about three in the afternoon, and proceeded at once to make the experiments. But now, an important difficulty presented itself. We had no dog.[18]

In 1750, long before carbon dioxide and oxygen were known to chemists, the Abbé Nollet, FRS, communicated his observations to the Royal Society of London. He provided a thorough physical description of the cave and how he spent much of the day extinguishing flaming torches in it, witnessing the near asphyxiation of a little dog, who recovered after passing out, and experimenting with other animals—a rooster died immediately, but several frogs, flies, a beetle, and some butterflies soon recovered. His scientific colleague, René-Antoine Ferchault de Réamur, had tried similar tests with a toad, a lizard, and a large grasshopper, all of which showed remarkable resistance to the cave vapors. Emboldened, Nollet plunged his own head into the cave (holding his breath) and concluded that the airs themselves were not poisonous or mephitic to animals, but acted by "reducing them to the impossibility of breathing their proper element."[19]

Alfred S. Taylor, a lecturer on medical jurisprudence in Guy's Hospital, provided a thorough chemical and geological analysis of the Grotto del Cane and its vapors.[20] He noted that most accounts were in guidebooks or by travelers, and were merely descriptive. Taylor thought that mephitic cave gas may be able to influence the body without being inhaled, that is, through the pores of the skin, or through absorption in the stomach since, as he wrote, "The intoxicating power of certain fluids impregnated by this gas, as well as the dizziness, vertigo, and other symptoms of cerebral disturbance . . . have long been known."[21] So, he concluded, such gas "has a distinctly deleterious influence upon the animal functions, in whatever way it may be introduced into the system."

Much later the chemist Sir Humphrey Davy—widely known for his practice of inhaling a large number of gases until nearly unconscious and then writing about the effects in his journal—explored phenomena attending suffocation by carbonic acid. He noted that aspiration of highly concentrated CO_2 triggered a spasmodic closing of his glottis and, being an acid, a strong, pungent sensation in his mouth. He noted this involuntary reaction prevented him from "taking a single particle of carbonic acid into my lungs." Later he was able to breathe a 30 percent concentration of CO_2 for about a minute, producing a slight sensation of giddiness and an inclination to sleep.[22]

In 1828–29 Colonel William Monteith of the Madras Engineers explored Azerbaijan and the shores of the Caspian Sea, visiting the celebrated Cave of Secundereah near Bosmitch, a grotto known to be filled with toxic vapors. After testing the air by lighting a brush fire, Monteith and his party descended cautiously into the cavern where they found themselves standing on animal bones—dog, deer, fox, and wolf. The torches they tossed into the gloom soon flickered out, and the hens they had brought along died in half a minute. Monteith apparently had a very good time in the cave, sprinkling naphtha on the fires, lighting fuses (which burned), and detonating several pounds of gunpowder. That act, "repeated often, had the effect of so entirely filling the cave with smoke, that [they] could no longer see anything at the bottom." During a subsequent visit, in the winter after a strong gale had passed, Monteith was able to descend much further into the cave than before. He speculated that the atmospheric conditions had served to ventilate the noxious gases.[23] The general inaccessibility of the site, however, kept the Cave of Secundereah off the Grand Tour.

The Valley of Death in Batur, Java, also known in Javanese as the *Guevo Upas*, lies in the shadow of an active volcano. It is, according to *Chambers's Edinburgh Journal*, "a great natural sepulcher where no bird can alight, nor beast stray, nor human set foot, and live."[24] In 1831 Alexander Loudon and his party visited the site, and after a bit of scrambling and a whiff of a "strong nauseous, sickening, and suffocating smell, "were lost in astonishment at the awful scene below . . . [the valley floor] covered with the skeletons of human beings, tigers, pigs, deer, peacocks, and all sorts of beasts and birds."[25] Loudon's party, with lighted cigars to warn them, descended the sides of the valley to within eighteen feet of the bottom, at which point they "sent in" a small dog attached to an eighteen-foot bamboo pole. "We had our watches in our hands, and in fourteen seconds he fell on his back; he did not move his limbs or look round, but continued to breathe eighteen minutes." A second dog, rushing to the side of his distressed companion, swooned in ten seconds and died in seven minutes. A fowl died in ninety seconds and another died before landing on the valley floor. The bleached skeleton of a man, said to have been a rebel seeking refuge, lay just out of reach. So intense was the carnage and so interested were the observers in the "awful scene" that they disregarded a heavy downpour and had to scramble for their lives in the mud to climb back out of the toxic valley. Legend attributed the mortality to a mythical "hydra tree of death." Loudon did not speculate on the causes, but in 1837 W. H. Sykes placed the blame squarely on emissions of carbon dioxide from the volcanic crater. Toxic events in volcanic regions are not at all rare. Recently, the *Jakarta Globe* reported that over one thousand people living on the slopes of the volcanic Mount Dieng near Batur were evacuated as a precaution due to fears of carbon dioxide and carbon monoxide poisoning. A nearby sign warns, "Alert!!! Area Prone to Toxic Gas." In 1979, 149 people were suffocated by emissions from the volcano.[26]

America has its toxic landforms too. In 1888, Walter Harvey Weed of the US Geological Survey reported a "Death Gulch" in the northeast corner of Yellowstone National Park. The dark and gloomy gulch, which he compared to Java's Valley of Death, contained the bodies of two bears and the skeletons of four more, along with the bones of elk, squirrel, hares, and the bodies of numerous insects.[27] Weed's description juxtaposed the natural beauty of the volcanic landscape with the grisly scenes of sudden death by asphyxiation. Death Gulch soon became a tourist attraction. T. A. Jaggar visited the gulch with a geological party in 1897 and published some photographs with his account. He found the remains of eight bears, one recently deceased, with no external signs of violence on the carcass.[28] In 1904 F. W. Traphagen of Montana State College transported chemical analysis equipment to the site, finding a carbon dioxide concentration of 10 percent in the gulch and 50 percent in some crevices. He suspected the animals died of sulphuretted hydrogen poisoning combined with the effects of carbon dioxide.[29] While there have been no recorded deaths of humans at the park, animal casualties due to toxic gas emissions continue to occur. The National Park Service attributed the mysterious deaths of five otherwise healthy bison in 2004 at

Norris Geyser Basin to deadly emissions of hydrogen sulfide and/or carbon dioxide gases.[30]

Although toxic levels of carbon dioxide at Yellowstone National Park have yet to cause any known human casualties, the same cannot be said for Mammoth Mountain, California, site of a well-known ski area. The mountain's toxic reputation emerged following a series of earthquakes in 1990 when a forest ranger was almost asphyxiated in a snow-covered cabin. He survived, but massive numbers of trees were dying on the mountain where soil carbon dioxide levels exceeded 30 percent.[31] When snow falls on the mountain, carbon dioxide emitted by fumaroles collects in low-lying, poorly ventilated areas, potentially reaching lethal levels.

Jean Ray says she clearly remembers the crisp blue sky of May 24, 1998, when she dropped off her husband Donald at Mammoth Mountain for a Sunday of cross-country skiing around Horseshoe Lake. It was the last time she saw him alive. He was found the next morning, skis on, in a snow cave beside an outbuilding. Carbon dioxide poisoning was suspected, but the coroner's report was inconclusive even though the concentration at the site was 70 percent. Unlike carbon monoxide, which bonds with hemoglobin and turns skin and mucous membranes cherry-red, CO_2 asphyxiation leaves few traces.[32] Where trails pass near volcanic vents, safety fences warn skiers away. But after a heavy snow in spring 2006, two members of the ski patrol maintaining the fences fell into a 21-foot-deep cavern filled with carbon dioxide. Both died, as did one of their intended rescuers.[33]

ALCHEMISTS AND EARLY CHEMISTS

Johann Baptista van Helmont (1577–1644) was both an alchemist and a chemist—a philosopher of fire, as he called himself. He was widely known for his great learning, careful observations, meticulous experiments, and work ethic. It was said he hardly ever stirred out of doors and was scarcely known by his neighbors. Yet he was a generous man who provided funds for the poor. One day, van Helmont burned sixty-two pounds of charcoal in a closed vessel. One pound of ashes remained, along with sixty-one pounds of an invisible substance that tried to escape its container. He coined the term "gas" for this substance and called it, variously, *spiritus sylvestre, gas carbonum, gas vinorum, gas uvarum,* and *gas musti.* His favorite term for the gas was "wild spirit," because it was difficult to collect, and because he believed it contained the essence of the charcoal that was seeking to escape. Van Helmont found "wild spirit" in caves, cellars, and the poisonous gas of mines; it sparkled in the bubbles of spa water and was produced by fermentation of grape juice and fermentation in the gut; it was produced when crab stones[34] were soaked in vinegar. In his medical studies, he discovered that "wild spirit" was condensed in blood and exhaled during respiration. Today we know this "wild spirit" of combustion, fermentation, and respiration as carbon dioxide, or CO_2.

The lethality of this gas appeared in the scientific literature in the 1750s when the Edinburgh physician and chemist Joseph Black (1728–99) generated mephitic air (an ancient alchemical term referring to either nitrogen or carbon dioxide and named for the vapors of the Temple of Mephitis) that asphyxiated birds and small animals. Black is credited as the founder of modern quantitative chemistry and as the discoverer of caloric principles now identified with latent and specific heat. His dissertation of 1754 (deemed "historic" by the *Dictionary of Scientific Biography*) was on "The Acid Humour Arising from Food and Magnesia Alba." Two years later Black published a significant expansion of this work called *Experiments Upon Magnesia Alba, Quicklime, and some other Alcaline Substances*. Here he identifies "fixed air" as an aeriform fluid and provides a quantitative measure of its presence in various minerals, including chalk and potash. Black further discovered that fixed air did not allow combustion, was fatal to animals, had a greater density than common air, was the product of animal respiration and the combustion of charcoal, and was given off in alcoholic fermentation. In many ways Black repeated, a century later, van Helmont's discoveries with fixed air instead of wild spirit.

Daniel Rutherford (1749–1819), a student of Black's, was Physician to the Royal Infirmary, Keeper of the Royal Botanic Garden in Edinburgh, King's Botanist in Scotland, and perhaps not coincidentally, since both were accomplished writers, the uncle of Sir Walter Scott.[35] In his medical dissertation of 1772, "Mephitic Air, which some call Fixed Air," he referred to "that singular species of air which is fatal to animals, which extinguishes fire and flame, and which is attracted with great avidity by quick-lime and alkaline salts." He set out to investigate the fixed air "in other things, animal, vegetable, or mineral [that] seem necessary to their constitution." He kept a mouse in a confined space until it died; then he burned, sequentially, a candle and some phosphorus in the remaining air until they too were extinguished. Using caustic lye to remove all the mephitic air (CO_2) from the container, he discovered that "what remains does not become in any way more wholesome; for although it produces no precipitate in lime-water, it extinguishes both flame and life no less than before." Rutherford had succeeded in isolating another asphyxiant gas: nitrogen, a substance he called noxious or phlogisticated air.[36]

The molecule CO_2 received its modern name from Antoine-Laurent Lavoisier (1743–94), who opened his landmark book on chemical nomenclature with the following profound thought: "We think only through the medium of words. Languages are true analytical methods . . . The art of reasoning is nothing more than a language well arranged."[37] Lavoisier divided atmospheric air into two elastic fluids, one fit for respiration and the other incapable of being respired. He identified fixed air or *gaz acid carbonique* (carbonic acid gas) as a combination of oxygen and carbon. Lavoisier wrote of the ubiquity and abundance of carbonic acid in nature, in chalk, marble, and limestone. Some sulfuric acid was all that was needed to disengage the carbonic acid and release it as a gas. He credited Black with "our first knowledge of [carbonic acid gas] . . . before whose time its property of remaining always in the state of gas had made it elude the researches of chemistry." According to Lavoisier, "this gas is incapable of being condensed into the solid or liquid form by a degree of cold or pressure hitherto known."[38] This was true enough at the time, but only for about three decades.

Michael Faraday (1791–1869), best known for his contributions to electromagnetism, became the first to liquefy carbonic acid in 1823 by cooling and pressurizing the gas in a glass tube.[39] This work was extremely dangerous, since the tubes were apt to explode. He described the liquid as "a limpid colorless body, extremely fluid, and floating upon the other contents of the tube," with a vapor pressure of 36 atmospheres at a temperature between 32 and 0°F.[40] Faraday's technique for liquefying CO_2 involved a closed glass tube filled with carbonic acid gas, heated at one end and cooled at the other in an ice bath. When the pressure reached 36 atmospheres, the gas condensed into a liquid in the chilled side of the tube. Such high pressures caused a lot of glassware to shatter, and in Faraday's laboratory, the air was often filled with shards of flying glass.

When, in 1834, Adrien-Jean-Pierre Thilorier (1790-1844) opened the stopcock of a pressurized container of liquid carbonic acid, a flurry of crystals of solid CO_2 appeared like a fine snow. As the carbon dioxide escaped, there was "un véritable et abondante pluie de neige." Thilorier (who is described as being a mechanician rather than a physicist, and still less a chemist) explained this phenomenon to a special commission of the French Academy as being the result of the solidification of water vapor out of the air. The commissioners were not satisfied, however, since the quantity of snow seemed large compared to the amount of water vapor in the air. Later it was determined that the rapid cooling by expansion had converted liquid carbon dioxide into a solid. Thilorier reported, "I had the honor, at the last session, to state to the Academy the phenomena which accompany the lique-faction of carbonic acid gas: I now announce the fact, important to science, of the solidification of this gas. This [is the] first instance of a *gas becoming solid and concrete.*" [41] Faraday called this "one of the most beautiful experi-mental results of modern times."[42]

At least one fatality is attributed to a dry ice explosion. On December 30, 1840, Thilorier, his assistant Osmin Hervy, and one other helper were preparing the metallic apparatus for producing carbonic acid gas, liquefy-ing it at high pressure (150 atmospheres), and generating solid carbon diox-ide "snow" in a rapid expansion. The demonstration was for the benefit of students in the Parisian School of Pharmacy. Thilorier had performed this experiment all over Europe more than 500 times—just two days earlier, to 1,200 students—without the slightest accident, but on this fateful day, he had just vacated the room when the pressurized apparatus gave way with a detonation like a bombshell. The explosion drove Hervy through a closet door, breaking both of his legs and causing severe bodily injuries. The other assistant was thrown prostrate on the floor but was otherwise uninjured. Everything in the laboratory was destroyed. Hervy was instantly removed to La Pitié-Salpêtrière Hospital where one of his legs was amputated, but he died from his injuries four days later. This unfortunate accident marked the end of public demonstrations on liquid and solid carbon dioxide in Paris.[43]

So-called dry ice made it possible to liquefy other gases and conduct ex-periments at low temperatures. The Liquid Carbonic Company, established in 1888 in Chicago, specialized in the production and delivery of cylinders of compressed gas for use in industry. Carbon dioxide was widely used to pressurize and dispense beer from the cellars of pubs, carbonate it in the bottle, and even chill it. A plethora of industrial applications followed. As funeral directors developed into their twentieth-century role of "custodians for the dead," the industry increasingly used dry ice to preserve cadavers, especially for shipping and for storage in the winter in cold climates when the ground was frozen solid. Bodies were packed in solid carbon dioxide at a temperature of $-78.5°C$, a technique that temporarily delayed the onset of decomposition. Dry ice is an effective preservative since it does not melt (it really is dry). It disappears directly into vapor, a process called sublimation. In the 1940s the Dottridge Brothers of London marketed 25-pound cylinders of Drikold, a dry ice product of Imperial Chemical Industries.[44] Today, Continental Carbonic Products advertises a kind of inexpensive "green burial" using dry ice (instead of embalming fluids) to preserve the human body for the funeral, with subsequent burial in a cloth shroud or simple biodegradable casket (instead of a metal one in a concrete vault) in a cemetery more likely to be filled with wildflowers than with stone monuments.[45]

Carbon Dioxide, "Climate Killer"

The ubiquitous carbon dioxide molecule is a good absorber and emitter of infrared radiation. Its heating effect is magnified by the presence of a much more abundant radiatively active gas, water vapor. These gases raise Earth's average surface temperature by 33°C. Anthropogenic activities such as fuel burning and deforestation are increasing the atmospheric content of carbon dioxide to the 0.04-percent level. Physical theory indicates that a doubling of the CO_2 concentration would heat the Earth 1.2°C, but water vapor feedbacks may add a factor of three or four to this, making it a potential "climate killer," at least for civilization as we know it.[46] The stories of the discovery of carbon dioxide as a powerful agent of climate change by nineteenth- and early twentieth-century scientists such as John Tyndall, Svante Arrhenius, Nils Ekholm, and Guy Stewart Callendar have been told elsewhere.[47] They lived prior to the current era of climate anxiety, which dates to 1988 when scientist James Hansen of the National Aeronautics and Space Administration announced to the US Congress and the world that, at least in his opinion, "Global warming has begun."[48] Since then, apprehensions have been multiplying rapidly that we are approaching a crisis in our relationship with nature, one that could have potentially catastrophic results for the sustainability of civilization and even the habitability of the planet. This fear is driven by the apprehension that Earth may experience a sudden and possibly catastrophic warming caused by industrial pollution. By 2006 a Nobel Laureate in chemistry had offered a "modest proposal" to cool the planet by bombarding the stratosphere with sulfate aerosols, thus cutting off solar insolation, but turning the blue sky permanently milky white.[49] Others proposed fantastic and unworkable techniques to capture and sequester CO_2 in concentrated amounts in underground storage caverns and in the depths of the sea.[50] If this is to be done, CO_2 must be stored in perpetuity—or else!

An unexpected fatal "burp" of naturally occurring carbon dioxide in an otherwise bucolic valley points to the dangers of underground storage. In 1986 at least 1,700 people lost their lives from a sudden and catastrophic release of carbon dioxide from the waters of Lake Nyos in Cameroon, West Africa. The worst such disaster in recent history came upon the villagers without warning when, apparently, an underwater rockslide agitated a huge pool of dissolved gas in the depths of the crater lake.[51] According to survivors, late in the evening of August 21 a deep rumbling echoed across the lake, followed by bubbling sounds, a plume of white vapor, and a surge of water accompanied by a sulfurous odor, although this latter sensation could have been a hallucination due to the anesthetizing effects of carbon dioxide. The luckier ones lost consciousness soon thereafter, awakening some six to thirty-six hours later, weak and confused, to find themselves surrounded by family and neighbors asphyxiated by the cloud. Lamps, full of oil, had been extinguished; pets and livestock were dead; and the entire region was eerily silent, devoid of birds for several days.[52] Two years earlier thirty-seven people had died from a similar, if smaller, outgassing at nearby Lake Monoun. Some interpreted these events as the revenge of the lake gods; others suspected, since there was no property damage, that it was a test of a neutron bomb. Geologists say that the crater lake had been accumulating dissolved carbon dioxide at supersaturated levels in its depths prior to the killing eruption. Today pipes installed in both lakes serve as release valves, venting the high-pressure gas high into the air like fizzy fountains.[53]

Natural disasters such as these argue for extreme caution by those planning to store carbon dioxide at high concentrations and pressures in old oil and gas fields, in the oceans, and in other geological formations. Some geoengineers are proposing to capture CO_2 from the air, liquefy it, and sequester it by injecting it into the depths of the sea—a massive engineering task that may be both economically and thermodynamically impossible—by some estimates it would take a 30 percent expansion of the world's energy use just to do this. Carbon sequestration is not simple; according to geologist William Evans, "any time you change the fluid pressure within the Earth, you risk creating seismicity. You have to be very careful how you do this."[54] The long-term storage of captured carbon dioxide is, like storing radioactive waste, a high-tech, risky, contentious, yet poorly understood undertaking.[55] A recent study suggests that carbon dioxide gas injection is not wholly benign and may have triggered earthquakes of magnitude three and larger in the Cogdell oil field of Texas.[56]

In a review article on the dynamics of lake eruptions and possible ocean eruptions, geoscientists Youxue Zhang and George King warn about possible instabilities involving attempts at carbon sequestration (recall the explosion that killed Thilorier's assistant in 1840): "If the amount of gas is small and not concentrated, bubbles would rise and dissolve in seawater. If a large amount of concentrated CO_2 gas is formed at . . . depth, the gas-bearing water might rise as a CO_2 bubble plume to reach the atmosphere, and thus defeat the purpose of carbon sequestration. Furthermore, the CO_2 bubble plume might erupt violently if the amount of CO_2 is high," similar to an eruption of lakes Nyos and Monoun. In a model of understatement they conclude, "In designing injection schemes, it is important to avoid such consequences."[57] Erupting plumes of lethal vapors may be the future of CO_2, especially if plans for its massive sequestration underground somehow go awry.

Wild ideas about fixing the sky abound today. Ocean iron fertilizers propose adding iron to the oceans to sequester CO_2. Klaus Lackner of the Earth Institute at Columbia University envisions a world filled with millions of inverse chimneys, some of them more than 300 feet high and 30 feet in diameter, inhaling up to 30 billion tons of carbon dioxide from the atmosphere every year (the world's annual emissions) and sequestering it in underground or undersea storage areas. Lackner refers to his scrubbers as "artificial trees," covered in CO_2-absorbing artificial leaves. Note, however, that trees sequester carbon, not carbon dioxide, and they return oxygen to the atmosphere. Unlike Lackner's forests, natural trees also provide shade, habitat, and food for squirrels, birds, and other living things.[58] Lackner's scheme is highly improbable, prohibitively expensive, and entails immense storage problems, since CO_2 is still *spiritus letalis*.

Perhaps the industrial waste stream and carbon dioxide's implication in anthropogenic climate change has completely overshadowed its earlier history. Perhaps not. With environmental concerns foremost, as Earth both metaphorically and literally runs out of breath,[59] and as climate engineering schemes take aim at capturing and sequestering fixed air or otherwise fixing the sky, carbon dioxide richly deserves its own molecular biography, dedicated to challenging the dominant reductionist and deterministic paradigm wherever it exists, at 4 percent or 0.04 percent concentrations.

NOTES

1. Soentgen, "On the History."

2. Moore, *New Air World*, 35–37.

3. National Institute of Occupational Safety and Health, http://www.cdc.gov/niosh/.

4. Fleming, *Historical Perspectives*, ch. 6; Fleming, *Callendar Effect*, 69.

5. Quinn and Jones, *Carbon Dioxide*.

6. Mick Kelley, personal communication.

7. Pausanias, *Description of Greece*, bk. 10, ch. 24.

8. Plutarch, *De Defectu*.

9. Plutarch, *Morals*, 475.

10. Pausanias, *Description of Greece*, bk. 9, ch. 39.

11. Seontgen, "On the History," 141.

12. Pliny the Elder, *The Natural History*, trans. John Bostock and H. T. Riley (London: Henry G. Bohn, 1855), bk. ii, ch. 95, http://www.perseus.tufts.edu/hopper/text?doc=Plin.+Nat.+toc&redirect=true; Dunglison, *Medical Lexicon*, 808.

13. Fox, "Historical and Scientific Aspects."

14. Dis was ruler of the underworld; Acheron the river of pain. Virgil, *Aeneid*, vii, 563–71.

15. Virgil, *Aeneid*, vi, 237–42.

16. Author's liberal interpretation of a poem attributed to "Genoino."

17. Greenwood, *Haps and Mishaps*, 343–44.

18. Dumas, *Impressions de voyage*, 62–70; Twain, *Innocents Abroad*, 322.

19. Nollet, *Extract of Observations*.

20. Taylor, "Account of the Grotta del Cane."

21. Ibid., 284–85.

22. Davy, *Collected Works*, vol. 3, 281.

23. Monteith, "Journal of a Tour," MSS Eur B 24 and B 25, India Office, British Library.

24. "Valley of Poison in Java."

25. Sykes, "Remarks on the Origin"; Loudon, "Extract of a Letter."

26. Elisabeth Oktofani, "Alert Raised over Volcano's Deadly Gas," *Jakarta Globe*, May 31, 2011, accessed Apr. 14, 2012, http://www.thejakartaglobe.com/home/alert-raised-over-volcanos-deadly-gas/444182.

27. Weed, "Deadly Gas-Spring."

28. Jaggar, "Death Gulch."

29. Traphagen, "Death Gulch."

30. Heasler and Jaworowski, "Geologic Overview."

31. Laura Hainsworth, "Solving the Mammoth Mountain CO_2 Mystery," *Science and Technology Review*, June 1996, accessed Jan. 19, 2012, https://www.llnl.gov/str/pdfs/06_96.3.pdf; US Geological Survey, *Invisible CO_2 Gas Killing Trees at Mammoth Mountain, California*, 2001, accessed Apr. 22, 2012, http://pubs.usgs.gov/fs/fs172–96.

32. Kevin Couglin, "Death of a Skier Points to an Invisible Danger," *Newark Star Ledger*, Mar. 6, 2005, accessed Apr. 15, 2012, http://www.nj.com/business/energy/ledger/index.ssf?/business/ledger/stories/0306deathskier.html.

33. Steve Hymon and Amanda Covarrubias, "Mammoth Ski Patrol Deaths Hit

Swiftly," *Los Angeles Times*, Apr. 8, 2006, accessed Apr. 22, 2012, http://articles .latimes.com/2006/feb/06/local/me-mammoth6.

34. Or *gastroliths*—calcareous matter from the stomachs of European crawfishes used as an antacid.

35. Charles D. Waterston and A. Macmillan Shearer, *Former Fellows of the Royal Society of Edinburgh 1783-2002: Biographical Index* II (Edinburgh: Royal Society of Edinburgh, 2006), accessed Feb. 8, 2011, http://www.rse.org.uk/fellowship/ fells_indexp2.pdf.

36. Rutherford, "On the Air," 372.

37. Lavoisier, *Elements of Chemistry*, citing the Abbé de Condillac, *System of Logic*.

38. Ibid., 229–30.

39. Steinle, "Michael Faraday."

40. Faraday, *Experimental Researches*, 92, 108; Faraday, "Historical Statement."

41. Thilorier, "Solidification"; Thilorier, "Lettre"; Roller, "Thilorier."

42. Almqvist, *History*, 96.

43. "Solidification of Carbonic Acid Gas."

44. Brian Parsons, "Change and Developments in the British Funeral Industry During the 20th Century, with Special Reference to the Period 1960–1994," PhD diss., University of Westminster, 1997, accessed Apr. 15, 2012, http://westminster research.wmin.ac.uk/8561/1/Parsons.pdf; Parsons, "Unknown Undertaking: The History of Dottridge Bros., Wholesale Suppliers to the Funeral Industry," n.p., accessed Apr. 15, 2012, http://www.bifd.org.uk/article_dottridge_june09.pdf.

45. Continental Carbonic Products, accessed Apr. 15, 2012, http://www.dryice source.com/dryice/daisies.php.

46. Soentgen, "On the History."

47. Fleming, *Historical Perspectives*; Crawford, *Arrhenius*, 74–82; Fleming, *Callendar Effect*.

48. "Global Warming Has Begun."

49. Crutzen, "Albedo Enhancement."

50. Fleming, *Fixing the Sky*, 225ff.

51. Sager, "Silent Death."

52. Kling et al., "1986 Lake Nyos," 169.

53. Kling et al., "Degassing Lakes Nyos and Monoun."

54. Couglin, "Death of a Skier."

55. Toth, *Geological Disposal*.

56. Wei Gan and Cliff Frohlich, "Gas Injection May Have Triggered Earthquakes in the Cogdell Oil Field, Texas," *Proceedings of the National Academy of Sciences*, early ed., www.pnas.org/cgi/doi/10.1073/pnas.1311316110.

57. Zhang and King, "Dynamics of Lake Eruptions," 314.

58. Fleming, *Fixing the Sky*, 250–51.

59. Nagy, "As the World."

REFERENCES

Almqvist, Ebbe. *History of Industrial Gases*. New York: Kluwer, 2003.

Black, Joseph. *Experiments Upon Magnesia Alba, Quicklime, and some other Alcaline Substances*. Edinburgh: G. Hamilton and J. Balfour, 1756.

Crawford, Elisabeth. *Arrhenius: From Ionic Theory to the Greenhouse Effect*. Canton, MA: Science History Publications, 1996.

Crutzen, Paul J. "Albedo Enhancement by Stratospheric Sulfur Injections: A Contribution to Resolve a Policy Dilemma? An Editorial Essay." *Climatic Change* 77 (2006): 211–20.

Davy, Humphrey. *The Collected Works of Humphrey Davy*. Vol. 3, *Researches, Chiefly Concerning Nitrous Oxide*. Edited by John Davy. London: Smith, Elder, 1839.

Dumas, Alexandre. *Impressions de voyage: Le corricolo* [1851]. Paris: Michael Lévy Frères, 1872.

Dunglison, Robley. *Medical Lexicon, A Dictionary of Medical Science*. Philadelphia: Blanchard and Lea, 1851.

Faraday, Michael. *Experimental Researches in Chemistry and Physics*. London: Taylor and Francis, 1859.

———. "Historical Statement Respecting the Liquefaction of Gases." *Quarterly Journal of Science, Literature, and the Arts* 16 (1823): 229–40.

Fleming, James Rodger. *Fixing the Sky: The Checkered History of Weather and Climate Control*. New York: Columbia University Press, 2010.

———. *Historical Perspectives on Climate Change*. New York: Oxford University Press, 1998.

———. *The Callendar Effect: The Life and Work of Guy Stewart Callendar (1898–1964), the Scientist Who Established the Carbon Dioxide Theory of Climate Change*. Boston: AMS Books, 2007.

Fox, Denis L. "Some Historical and Scientific Aspects of Narcosis, with Special Reference to Carbon Dioxide as a Narcotic." *Medical Review of Reviews* 38, no. 9 (1932): 515–42.

"Global Warming Has Begun, Expert Tells Senate." *New York Times*, June 24, 1988, 1.

Greenwood, Grace. *Haps and Mishaps of a Tour in Europe*. Boston: Ticknor, Reed and Fields, 1854.

Heasler, Hank, and Cheryl Jaworowski, "Geologic Overview of a Bison-Carcass Site at Norris Geyser Basin." Yellowstone National Park: US Park Service, 2004.

Jaggar, T. A., Jr., "Death Gulch: A Natural Bear-Trap." *Popular Science Monthly* 54, no. 4 (1899): 475–81.

Kling, George W., Michael A. Clark, Harry R. Compton, Joseph D. Devine, William C. Evans, Alan M. Humphrey, Edward J. Koenigsberg, John P. Lockwood, Michele L. Tuttle, and Glen N. Wagner. "The 1986 Lake Nyos Gas Disaster in Cameroon, West Africa." *Science* n.s. 236 (1987): 169–75.

Kling, George W., William C. Evans, Greg Tanyileke, Minoru Kusakabe, Takeshi Ohba, Yutaka Yoshida, and Joseph V. Hell. "Degassing Lakes Nyos and Monoun: Defusing Certain Disaster." *Proceedings of the National Academy of Sciences* 102, no. 40 (2005): 14185–90.

Lavoisier, Antoine-Laurent. *Elements of Chemistry in a new systematic order, containing all the modern discoveries* [1789]. Translated by Robert Kerr. Edinburgh: William Creech, 1790.

Loudon, Alexander. "Extract of a Letter from Mr. Alexander Loudon to W. T. Money, Esq. 24 May 1831 . . ." *Nautical Magazine and Naval Chronicle* (London 1841): 380–82.

Monteith, [William]. "Journal of a Tour through Azerdbijan and the Shores of the Caspian." *Journal of the Royal Geographical Society of London* 3 (1833): 1–58.

Moore, Willis Luther. *The New Air World: The Science of Meteorology Simplified.* Boston: Little, Brown, 1922.

Nagy, Gregory. "As the World Runs Out of Breath: Metaphorical Perspectives on the Heavens and the Atmosphere in the Ancient World." In *Earth, Air, Fire, Water: Humanistic Studies of the Environment*, edited by J. C. Ker Conway, K. Keniston, and L. Marx, 37–50. Amherst: University of Massachusetts Press, 1999.

Nollet, Abbé. "Extract of the Observations made in Italy, by the Abbé Nollet, FRS on the Grotta de Cani." Translated by Tho. Stack, MD, FRS. *Philosophical Transactions (1683–1775)* 47 (1751–1752): 48–61.

Pausanias. *Description of Greece.* Translated by W. H. S. Jones. Loeb Classical Library. Cambridge, MA: Harvard University Press, 1918.

Plutarch. *De Defectu Oraculorum.* Loeb Classical Library. Cambridge, MA: Harvard University Press, 1936.

———. *Morals*, Vol. 5. Boston: Little, Brown, and Co., 1878.

Quinn, Elton L., and Charles L. Jones. *Carbon Dioxide.* New York: Reinhold, 1936.

Roller, Duane D. "Thilorier and the First Solidification of a 'Permanent' Gas." *Isis* 43 (1951): 109–13.

Rutherford, Daniel. "On the Air Called Fixed or Mephitic" [1772]. Translated by C. Brown. *Journal of Chemical Education* 12 (1935): 370–75.

Sager, Curt. "Silent Death from Cameroon's Killer Lake." *National Geographic* (Sept. 1987): 404–20.

Soentgen, Jens. "On the History and Prehistory of CO$_2$." *Foundations of Chemistry* 12 (2010): 137–48.

"Solidification of Carbonic Acid Gas. Explosion of the Apparatus. Death of M. Hervy." *Provincial Medical and Surgical Journal* 1, no. 16 (Jan. 16, 1841): 271.

Steinle, F. "Michael Faraday." *Dictionary of Scientific Biography.* Vol. 21. Detroit: Scribner's, 2008.

Sykes, W. H. "Remarks on the Origin of the Popular Belief in the *Upas,* or Poison Tree of Java." *Journal of the Royal Asiatic Society of Great Britain and Ireland* 4 (1837): 194–99.

Taylor, Alfred S. "An Account of the Grotta del Cane; with Remarks on Suffocation by Carbonic Acid." *London Medical and Physical Journal* n.s. 12 (1832): 278–85.

Thilorier, Adrien. "Lettre de M. Thilorier sur l'acide carbonique solide." *Comptes rendus* 3 (1836): 432–34. English translation in *American Journal of Science and Arts* 31 (1837), 31: 163, 404–405.

———. "Solidification de l'acide carbonique." *Comptes rendus* 1 (1835): 194–96.

Toth, Ferenc L., ed. *Geological Disposal of Carbon Dioxide and Radioactive Waste: A Comparative Assessment.* Advances in Global Change Research 44. Dordrecht: Springer, 2011.

Traphagen, F. W. "Death Gulch." *Science* n.s. 19, no. 485 (1904): 632–34.

Twain, Mark. *The Innocents Abroad* [1850]. Hartford: American Publishing Co., 1869.

"The Valley of Poison in Java." *Chambers's Edinburgh Journal* 8, no. 370 (Mar. 2, 1839): 46–47.

Virgil. *Aeneid.*

Weed, Walter H. "A Deadly Gas-Spring in the Yellowstone Park." *Science* n.s. 13, no. 315 (1889): 130–32.

Zhang, Youxue, and George W. King. "Dynamics of Lake Eruptions and Possible Ocean Eruptions." *Annual Review of Earth and Planetary Sciences* 34 (2006): 293–324.

CONTRIBUTORS

PETER BRIMBLECOMBE is professor of atmospheric environment at City University of Hong Kong and senior editor of *Atmospheric Environment*. He is the author of *The Big Smoke: A History of Air Pollution in London since Medieval Times* (Methuen, 1987). Additional details are available at http://www.uea.ac.uk/~e490/www.htm.

MATTHIAS DÖRRIES is professor for history of science and director of IRIST at the University of Strasbourg, France. His research and interests focus on French history of science and the geophysical and atmospheric sciences, particularly climate change and geoengineering. His most recent research includes articles on the history of volcanism and climate change, climate catastrophes and fear, and the nuclear winter theory.

RICHARD CHASE DUNN is a recent graduate in mechanical engineering from the University of South Carolina. His academic interests are geared toward the interaction between technology and environmental systems, particularly the automobile and the history of automobile engineering. As an undergraduate, he was awarded a Magellan Fellowship from the University of South Carolina to pursue research with Ann Johnson into the history of automobile emission control systems.

ROGER EARDLEY-PRYOR is a PhD candidate at the University of California, Santa Barbara, specializing in modern US history with concentrations in environmental history, the histories of science and technology, and world history. His dissertation, "The Global Environmental Moment: The Limits of Sovereignty and American Science on Spaceship Earth," examines American scientists' political activism in helping foster global environmentalism, particularly at the United Nations Conference on the Human Environment in Stockholm in 1972. Eardley-Pryor also researches environmental policies for nanotechnology and worked as a National Science Foundation research fellow at UCSB's Center for Nanotechnology in Society. He has taught courses at Washington State University–Vancouver and Portland State University.

JAMES RODGER FLEMING is a professor of science, technology, and society at Colby College and a visiting lecturer and officer at Columbia University in the City of New York. He has written extensively on the history of weather, climate, and environment. In 2010–2011 he was a Gordon Cain Fellow at the Chemical Heritage Foundation and convened the conference that led to this volume.

CHRISTOPHER HAMLIN is a professor in the department of history and the graduate program in history and philosophy of science at the University of Notre Dame, where he has been a member of the faculty since 1985. He is also an honorary professor in the department of public health and policy at the London School of Hygiene and Tropical Medicine. Hamlin has a long-standing interest in the history of environmental chemistry, chiefly as it affects health, and is the author of *A Science of Impurity: Water Analysis in Nineteenth-Century Britain* (University of California Press, 1990), dealing with the history of concepts of water quality, and *Public Health and Social Justice in the Age of Chadwick* (Cambridge University Press, 1998), dealing with the emergence of modern public health, and coauthor (with Philip T. Shepard) of *Deep Disagreement in U.S. Agriculture: Making Sense of Policy Conflict* (Westview, 1993), which applies interpretive social science to contemporary alternative agriculture policymaking, and *Cholera, the Biography* (Oxford University Press, 2009). He was a Cain Fellow at the Chemical Heritage Foundation in 2005–2006.

E. JERRY JESSEE is an assistant professor of history at the University of Wisconsin, Stevens Point. He is presently writing a book manuscript that explores the impact of scientific research on nuclear testing fallout on global ecological thinking and environmental science. The book is tentatively titled *Radiation Ecologies: Nuclear Weapons Testing, the Ecosphere, and the Rise of Environmental Science.*

ANN JOHNSON is an associate professor in the history and philosophy departments at the University of South Carolina. Her work concerns the history of applied science and engineering and especially science and engineering in the context of design. She is the author of *Hitting the Brakes: Engineering Design and the Production of Knowledge* and several articles on the history of nanotechnology. Her current projects focus on the changing role of mathematical and computational modeling in science and engineering from the Enlightenment to the present.

SUSIE KILSHAW is a principal research fellow in the department of anthropology at University College London. Broadly, her work focuses on the impact of culture on illness experience, primarily in Qatar and the United Kingdom. She has two Qatar-based research projects funded by the Qatar National Research Fund: one focuses on the public understanding of genetics and the other looks at women's experience of miscarriage. Her book, *Impotent Warriors: Gulf War Syndrome, Masculinity and Vulnerability* (Berghahn, 2009) investigates the emergence and construction of Gulf War syndrome (GWS) in the United Kingdom.

JONGMIN LEE is a lecturer in the department of engineering and society at the University of Virginia. He completed a PhD in STS from Virginia Tech in 2013. In his dissertation, "Engineering the Environment: Regulatory Engineering at the U.S. Environmental Protection Agency, 1970–1980," he examined a shift in the EPA's priorities from stringent health-based standards to flexible technology-based ones through the development of end-of-pipeline pollution control devices, which contributed to the emergence of economic incentives and voluntary management programs. His research interests include the potential and limitation of technology as a solution to environmental problems; regulatory governance in environmental and science policy; engineering studies; and public participation and democracy.

ANDREA POLLI is currently an associate professor in fine arts and engineering at the University of New Mexico in the area of Art & Ecology and the Mesa Del Sol Chair of Digital Media at the university. Polli's work with science, technology, and media has been presented widely in over one hundred presentations, exhibitions, and performances internationally, including the Whitney Museum of American Art Artport and the Field Museum of Natural History, and has been recognized by numerous grants, residencies, and awards including UNESCO. Her work has been reviewed by the *Los Angeles Times*, *Art in America*, *Art News*, *NY Arts*, and others. She has published several audio CDs, DVDs, two book chapters, and many papers with MIT Press, Cambridge University Press, and others.

RACHEL ROTHSCHILD is a PhD candidate in the history of science and medicine program at Yale University. Her academic interests include the history of earth and environmental science, international history, and environmental history in the twentieth century. She is writing her dissertation on the history of acid rain in Europe, examining the scientific studies and diplomatic negotiations undertaken to address the problem. Her dissertation project has received generous support from the American Meteorological Society, the American-Scandinavian Foundation, and the MacMillan Center for International and Area Studies. She is also the recipient of a National Science Foundation Graduate Research Fellowship, and intends to seek ways for her scholarship to find wider application among government officials and the public.

BRENDA GARDENOUR WALTER is an assistant professor of history at the Saint Louis College of Pharmacy, where she teaches courses in the medieval and modern history of medicine. Her research focuses on the intersection between medical and theological authority, and explores the ways that both of these authoritative languages were used to construct, inscribe, and interpret the human body, both real and imaginary. Other research interests include medieval cosmology and Aristotelian philosophy, postmodernism in the history of medicine, and the influence of nonsystematic medicine and theology—image, practice, and theory—on the modern horror genre.

INDEX

acid rain: Clean Air Act, 232; comparison to climate change, 105; EPA, 127; international regulation, 182–83, 191–93, 196–98; OECD study, xiii, 181–83, 192; sulfur dioxide, 234–35; Svante Odèn, 186–87

aether, 3–4

air injection reaction (AIR) systems (automobile), 115, 121

Airlight (artwork), 245–47

air quality management, 102, 104–5, 127, 245–47

alchemy, 259–60

American Chemical Society, 252

American Meteorological Society, 160

Anderson, Clinton, 164, 167–68

Anthropocene, 235

Aquinas, Thomas, 11, 15

Aristotle: on air, 2–5; on male and female bodies, 11–14

Arrhenius, Svante, 252, 263

art, 242–48

Atomic Energy Commission (AEC): attitudes towards nuclear fallout, 153–56; depletion of ozone layer, 218–20; fallout from Bikini Atoll test, 159; fallout from Nevada Test Site, 152–53; health effects of nuclear fallout, 156–59; High Altitude Sampling Project, 165–67; radioactive debris in the stratosphere, 164; radiotracers, 165–66

Austria, 188

automobiles: computerization, 122–24;

incentives to reduce emissions, 112–13, 120; smog, 109–11, 113; technologies to reduce emissions, 113–23, 210

Avicenna, 14

Balkin, Amy, 243–44

Barry, Robert, 242

Barth, Delbert, 132, 143

Bartlett, Elisha, 38–39

Batur, Java, 252, 257–58

Batzel, Roger E., 219

Beck, Ulrich, 87

Beckman, Arnold, 112

Belgium, 183, 188

Bellew, H. W., 40–43

Bernard, Claude, 27

Bikini Atoll, 159

biofuels, 103

Biological Effects of Atomic Radiation (BEAR) Committees, 159–60, 162

Black Death, 9. *See also* plague

Black, Joseph, 260–61

Blair, Henry, 152

bodies: effect of air on, x–xi; Gulf War syndrome, 77, 82, 84, 91; health scares, 90; in medieval thought, 5–7; World War I, 84

Boerhaave, Herman, 33

Boyle, Robert, 34

Brewer, Alan, 162

Brimblecombe, Peter, 109, 125n16, 225n3, 248n12

Britain, 183, 185, 189–90, 192, 196–97

Brodie, Benjamin, 33, 39

Brundtland, Gro Harlem, 192, 196

Bryden, James L., 40–43

Buisson, Henri, 210

Bundy, McGeorge, 53

California Institute of Technology, 97, 112, 210–11

Callendar, Guy Stewart, 263

Canada, 185, 194

carbon-14, 164

carbon dioxide: automotive emission, 110–11, 113, 115, 121, 123–24; California laws, 236; "cap and trade" system, 236–41, 243–44; in caves and valleys, xiii, 253–59; climate change, xiii, 123, 233, 263–65; environmental effects, 233; funeral arrangements, 262–63; as greenhouse gas, 232–33; history of, 259–62; in industry, 262–63; Kyoto Protocol, 236; liquid, 261–62; property rights, 238, 240; sequestration, 263–65; solid state of, 261–63; tax on emissions, 239; understandings of, xiii. *See also* "climate killer"; dry ice; fixed air; wild spirit.

carbon monoxide: automotive emission, 109–10, 113; EPA regulation, 131, 232; Kyoto Protocol, 236; technologies to reduce emissions, 115, 118–19, 121, 123–24

Carden-Coyne, Ana, 80

Cardinal Winds, 4–5, 8

Carson, Rachel, 59, 86

Castañega, Martin, 2

Castle Bravo thermonuclear test. *See* Bikini Atoll

catalytic afterburner (converter) (automobile), 119, 121–23, 143

Cave of the Dog. *See* Grotto del Cane

Cave of Secundereah, 257

Cave of Trophonius, 253

Central Electricity Generating Board (CEGB) (Britain), 197

Chandler, Raymond, 98

Chang, Julius S., 216, 220, 224

chemicals, toxic, 85–86

chemical weapons: Dugway Proving Ground accident, 59–60; Geneva Protocol, 52, 60–61, 63, 66; Gulf War syndrome, 77–81, 84, 89, 90; Khamisayah arms dump, 80–81, 89; lethal types, 51–52; nonlethal types, 50–52; Okinawa US Army base, 62; Operation CHASE, 61–62; Pugwash Conference, 53–54; soldiers' fear of, 80, 83–84; use in warfare, 51–56, 58–59, 83–84; World War I, 83–84

Chemical Warfare Service, US Army (CWS), 52

chemistry: automotive emissions, 109, 112, 116; carbon dioxide, 259–61; policy, 102–5, 110; smog, 95–105, 110, 112–13, 116–17

Chicago Climate Exchange (CCX), 237, 244

chlorofluorocarbons (CFCs): comparison to smog, 102; impact on ozone, 213, 234; policy, 208–9, 219, 232, 237–38

cholera: and the atmosphere, 23; contemporary work, 43; electrical cause of, 35; James L. Bryden and H. W. Bellew, 40–42; microbial cause of, 41–42; Reginald Orton, xi, 28, 31–35; Robert Koch, 41;

Chrysler Corporation, 118–19

Clean Air Act: automotive emissions, 115–16, 118, 121; "cap and trade" system, 237; environmental legislation, 184; EPA, 128, 131, 138; ozone, 212; summary of, 231–32

Clean Water Act of 1972, 128

climate change: and art, xiii, 230, 247–48;

El Niño/La Niña Southern Oscillation (ENSO), 43, 104

emission control, 101–2, 104, 127

emissions (automotive), xii; measurement, 116; ozone, 100–101; policy, 95, 100, 103, 210, 232; pollution, 109–111, 119; smog, 97–98, 109, 112, 123; technological fixes, 109–11, 113–24

Engebreth, Ben, 244–45

engine gas recirculation (EGR) system (automobile), 117–18, 121–22

environmentalism: anxieties about climate and health, 86–87; arguments against nonlethal chemical weapons, 51, 59

Environmental Protection Agency (EPA), xii; and automotive emissions, 118, 127; CHESS, 111, 127–29, 131–44; control technology, 143; coproduction of research and regulation, 141–42; creation of, 128–29; dual role of, 130, 142; National Environmental Research Centers (NERCs), 130–32, 140, 142; Office of Research and Development, 140, 142–43; Office of Research and Monitoring, 130–31, 142; and regulation, 130–32, 141–42, 232, 234; Science Advisory Board (SAB), 137–38, 140; separation of research and regulatory roles, 139–40

epidemiology. *See* disease causation

Eriksson, Erik, 186–89

Eudiometry, 30

European Air Chemistry Network, 186–87

Evans, William, 264

Eyring, Henry, 99

Fabry, Charles, 210

fallout (nuclear). *See* nuclear weapons; radioactivity

Faraday, Michael, 261–62

fever, thermic, 27, 39

Finklea, John, 132, 137–39, 143

Finland, 185

fixed air, 260–61

Fleming, James Rodger (Jim), 86, 211

Foley, Henry M., 215–17, 223

Foligno, Gentile da, 9

Ford, Gerald, 51, 66, 225

forest fires, 104

France, 184–85, 190

friendly fire, 88–90

Galen, 2, 7, 10–11, 14; Neo-Galenic, 33

Galston, Arthur, 65

Galvani, Luigi, 34

gasoline direct injection (GDI) (automobile), 122

General Mills, 164–65

General Motors, 117

Geneva Protocol: and chemical weapons, 52; and US debate over ratification of, 60–61, 63–67

Germany, 185

Giddens, Anthony, 87

Goethe, Johann Wolfgang von, 256

Gore, Al, 240

Grand Tour, 252, 255

Greater St. Louis Citizens' Committee on Nuclear Information (CNI), 169–70, 172

Greenfield, Stanley, 130–32, 138, 142–43

Greenfort, Tue, 243

Greenwood, Grace, 255

Gregory the Great (pope), 9, 16

Grotto del Cane, 252, 255–57

Guevo Upas, 257–58

Gulf War, 77–78, 80–83

Gulf War syndrome (GWS): biological and chemical weapons, 84, 90; biological and health anxieties, 86; and birth defects, 82–83; chemical sensitivity, 85–86; climate change, 86; contamination, 82; experiences of sufferers, 77, 79–82, 85–86, 91; explanations, xii, 78–82; friendly fire, 88–90; Khamisayah arms dump, 80–82, 89; oil well fires, 78–80; psychological aspects, 82; risk, 88, 90–91; World War I veterans, 83–84

Haagen-Smit, Arie Jan, 97–98, 101, 104, 112–13, 210
Haagen-Smit, Zus, 112
Hamlin, Christopher, 144n5
Hampson, John, 219, 224
Hansen, James, 86, 263
Harrison, Mark, 37
Heinen, Charles, 118–20
Helmont, Johann Baptista van, 259–60
Helsinki Protocol, 197
Hersch, Seymour, 60–62
Hervy, Osmin, 262
High Altitude Sampling Project (HASP), 165–67
Hippocrates, 2, 5, 7–10, 14; disease causation, 7–10; foundation for medieval medicine, 2; heritage of, 26, 28, 30–31, 39; Hippocratic tradition, 8–10; on menstrual blood, 14; Neo-Hippocratic, 26, 28, 30–31, 39; on the winds, 5
Hirsch, August, 43
Högström, Ulf, 189
Holinshed, Raphael, 1
Humboldt, Alexander von, 28, 36
humoral theory: and the cosmos, 5–7; and the environment, 7–9; male and female bodies, 11–16; and plague, 9–10
Hussein, Saddam, 78, 80
Huxham, John, 33
hydrocarbons: automotive emissions, 110;

EPA regulations, 131; smog, 98, 112–13, 210; technologies to reduce emissions, 113–19, 121, 123
hydrogen bomb (H-bomb), 156

Ibn-Jazzar, 11
Iklé, Fred C., 218–20, 223–24
India: cholera, 40–42; medical careers, 37–38; Reginald Orton, 31–34
Intergovernmental Panel on Climate Change (IPCC), 233
International Geophysical Year (IGY), 164–65, 188, 210
International Ozone Commission, 210
International Union of Geodesy and Geophysics, 210
iodine-131, 170–72
Isidore of Seville, 5–7, 9, 14
Italy, 185
Ivy-Mike test, 156–58

Jaggar, T. A., 258
James I of Scotland, 1–2
Jason Division, 214–15
Jessee, E. Jerry, 225n12
Johnson, James, 32, 36
Johnson, Lyndon B., 53–54, 56
Johnston, Harold S., 216–17, 222, 224
Joint Committee on Atomic Energy (JCAE), 162–64, 167–68

Katz, Morris, 99
Khamisayah arms dump, 80–82, 89
Kilshaw, Susie, 68n4
King, George, 264–65
Kissinger, Henry, 63
Klein, Yves, 242
Koch, Robert, 26, 28, 41
Kramer, Henry, 3, 15

National Oceanic and Atmospheric Administration (NOAA), 235–36
Netherlands, 188, 194
Neumann, John von, 211
Nevada Test Site (NTS), 152–53
Nider, Johannes, 3, 15
nitrogen, 231, 260
nitrogen oxides: acid rain, 191; and atmosphere, 231; automotive emissions, 110, 113; biofuels, 103; Clean Air Act, 232; EPA regulation, 131–32; international agreements, 197; Kyoto Protocol, 236; smog, 97, 210; technologies to reduce emissions, 110–11, 116–19, 121–23
Nixon, Richard: environmental agenda, 128, 225; EPA, 115; policy on chemical weapons, 62–64, 66–67; supersonic transport, 212
Nollet, Abbé, 256
Norris Geyser Basin, 258
Norway, 185, 188, 194–97
Norwegian Institute for Air Research (NILU), 187–91
nuclear material, 79, 81
nuclear weapons: and atmosphere, 173–74, 209; comparison to chemical weapons, 63; fallout, xii–xiii, 152–60, 169; health effects, 156–57, 171; hydrogen bomb, 156; Limited Test Ban Treaty, 172–73; nuclear winter, 222–25; ozone layer, 208–9, 213–22; science information movement, 169–70, 172

Occidens, 4–5
Odèn, Svante, 186–88
OH (hydroxyl) radical, 99–101
oil production: and fires, 78–80, 89; Gulf War syndrome, 79–80, 89; hydrocarbon emissions, 112–13; sequestration of carbon dioxide, 264
Operation CHASE, 61–62
Operation Ranger, 152–53
operations research, 135–36
Oracle at Delphi, 253
Orfila, Manuel, 30
Organisation for Economic Cooperation and Development (OECD): air pollution, xiii, 184–85, 187; long-range transboundary air pollution project, 181–83, 188, 190; pollution control, 191–95, 198
Organisation for European Economic Cooperation (OEEC), 181, 184
Oriens, 4–5
Orton, Reginald, 28–29, 31–32, 43; cholera, xi, 32–35; reception of his ideas, 23–24, 35–43
Ottar, Brynjulf, 187–90, 196–98
oxygen bars, 241–42
ozone: art, 246; automobile emissions, 100, 117; EPA, 131, 232, 234; health concerns, 36, 100–101, 117, 132–33; military uses, 211–12; properties, 234; scientific studies, 210; smog, 97–98, 209–10; and the stratosphere, 162, 209; perceptions of, xiii, 208–9, 212; volatile organic compounds, 100–101, 104, 236
ozone layer: changing views, xiii, 223; depletion, 86, 89, 102, 174, 208, 213–24, 232, 234, 237; hole, 105, 211; human modification, 211–12; location, 231; Montreal Protocol, 234, 237; nuclear weapons, xiii, 208–9, 213–22, 224; policy, 102, 105, 209, 232, 234; ultraviolet absorption, 208–10, 212–13, 234, 237; uncertainty, 220–24. *See also* supersonic transport

Siegelaub, Seth, 242

Silent Spring, 59, 86

Simms, Andrew, 240–41

smog: Arie Jan Haagen-Smit, 97–98, 101, 104, 112–13, 210; cause of disease, 36; chemistry, 97–101, 110–13, 116–17; emission control technologies, 111, 113; health concerns, 36, 100–101, 128; Los Angeles, xii, 95–98, 104–5, 112, 183–84, 209–10; ozone, 234; policy, 95, 100, 102–5, 110–11, 114–16, 121–22, 124; popular culture, 98; Raymond R. Tucker's report, 96–97; volatile organic compounds, 100–101, 103

Snow, John, 27

Society for Automotive Engineers (SAE), 114, 116, 118–19

soul, 2–3, 10–11

Southern California Gas Company, 97

spiritus letalis, 251, 254, 265

Sprenger, Jacob, 3, 15

Stern, Arthur C., 99

St. Louis, 96, 112, 183

Stockholm Declaration of 1972, 193–95

Stoermer, Eugene F., 235

stratosphere: human modification, 210–11; location, 231; nuclear weapons, 155, 158–64, 167–69, 213; supersonic transport, 212

strontium-90: fallout, 157–61; measurement, 163–65; policy, 167–73

student protests against Vietnam War, 64–65

Study of Critical Environmental Problems, 213

Study of Man's Impact on Climate, 213

sulfur dioxide: cap and trade system, 237–38; Clean Air Act, 232; decrease, 95; EPA, 131, 139; Kyoto Protocol, 236;

OECD, xiii, 183–89; policy, 192, 195–98; as pollutant, 234–35; smog, 97, 112

sunstroke, 27, 39

supersonic transport (SST): catalyst for stratospheric research, 212–13, 224; ozone layer, xiii, 208–9, 214–18, 222

Sweden, 185–86, 188

Switzerland, 185, 188

Sykes, W. H., 258

Syndenham, Thomas, 29–31, 33, 35, 43

systems theory, 130, 134–35

Taipei, 245–46

Taylor, Alfred S., 256–57

tear gas: arguments against, 51, 53–56, 62, 64, 65–66; arguments for, 50–51, 53–58, 64; CN (chloracetophenone) type, 51–52, 54; CS (o-chlorobenzylidene malononitrile) type, 52, 55–56, 58–59, 64; domestic use, 58–59, 62, 64–65; Geneva Protocol of 1925, 63–64, 66; paradox of, xi, 50; use in warfare, 51–56, 58–59, 61, 63–64, 66

technological fix, x; automotive emissions, xii, 109, 111, 113–18, 121–24; climate change, 240; tear gas, xi, 50–51, 58

Temple of Mephitis, 254

Tet Offensive, 58

tetra-ethyl lead, 119, 121

Thatcher, Margaret, 196–97

thermonuclear weapons. *See* nuclear weapons

thermostatic air cleaner (TAC) system (automobile), 115, 121

Thilorier, Adrien-Jean-Pierre, 261–62

toxic airs: antiquity, 8; carbon dioxide, 251–52; caves and valleys, 252–59; Clean Air Act, 232; as a concept, ix–xiv; EPA, 141; Middle Ages, 9–10; *pneuma,* 7, 89; St. Louis, 112; tailpipe emissions, 109;